Progress in Inflammation Research

Series Editor

Prof. Michael J. Parnham PhD
Senior Scientific Advisor
PLIVA Research Institute Ltd.
Prilaz baruna Filipovića 29
HR-10000 Zagreb
Croatia

Advisory Board

G. Z. Feuerstein (Wyeth Research, Collegeville, PA, USA)
M. Pairet (Boehringer Ingelheim Pharma KG, Biberach a. d. Riss, Germany)
W. van Eden (Universiteit Utrecht, Utrecht, The Netherlands)

Forthcoming titles:

Chemokine Biology: Basic Research and Clinical Appliation, Volume II: Pathophysiology of Chemokines, K. Neote, G.L. Letts, B. Moser (Editors), 2006
Endothelial Dysfunction and Inflammation, A. Graham (Editor), 2006
In vivo Models of Inflammation, 2nd edition, C. S. Stevenson, D. W. Morgan, L. A. Marshall (Editors), 2006
Lymphocyte Trafficking in Health and Disease, R. Badolato, S. Sozzani (Editors), 2006
The Resolution of Inflammation, A.G. Rossi, D.A. Sawatzky (Editors), 2006

(Already published titles see last page.)

The Hereditary Basis of Rheumatic Diseases

Rikard Holmdahl

Editor

Birkhäuser Verlag
Basel · Boston · Berlin

Editor

Rikard Holmdahl
Section for Medical Inflammation
Research
Department of Cell and Molecular
Biology
Lund University
Solvegatan 19, I11 BMC
S-221 84 Lund
Sweden

Library of Congress Cataloging-in-Publication Data

The hereditary basis of rheumatic diseases / Rikard Holmdahl, editor.
    p. ; cm. -- (Progress in inflammation research)
    Includes bibliographical references and index.
    ISBN-13: 978-3-7643-7201-9 (alk. paper)
    ISBN-10: 3-7643-7201-X (alk. paper)
    ISBN-13: 978-3-7643-7419-8 (e-ISBN)
    ISBN-10: 3-7643-7419-5 (e-ISBN)
    1. Rheumatism--Genetic aspects. I. Holmdahl, Rikard, 1953- II. PIR (Series)

    RC927.H47 2006
    616.7'23042--dc22

                                                   2006040762

Bibliographic information published by Die Deutsche Bibliothek
Die Deutsche Bibliothek lists this publication in the Deutsche Nationalbibliografie;
detailed bibliographic data is available in the internet at http://dnb.ddb.de

ISBN-10: 3-7643-7201-X Birkhäuser Verlag, Basel – Boston – Berlin
ISBN-13: 978-3-7643-7201-9 Birkhäuser Verlag, Basel – Boston – Berlin

Contents

List of contributors

Marta E. Alarcón-Riquelme, Department of Genetics and Pathology, Rudbeck Laboratory, Uppsala University, Dag Hammarskjölds väg 20, 751 85 Uppsala, Sweden; e-mail: marta.alarcon@genpat.uu.se

Anne Barton, Arthritis Research Campaign Epidemiology Unit, University of Manchester, Manchester, M13 9PT, UK

Gerd-R. Burmester, Department of Rheumatology and Clinical Immunology, Charité University Hospital, Schumannstrasse 20/21, 10117 Berlin, Germany; e-mail: gerd.burmester@charite.de

Peter K. Gregersen, Robert S. Boas Center for Genomics and Human Genetics, The Feinstein Institute for Medical Research, North Shore LIJ Health System, 350 Community Drive, Manhasset, New York 11030, USA; e-mail: peterg@nshs.edu

Joachim Grün, oligene GmbH, Schumannstrasse 20/21, 10117 Berlin, Germany; e-mail: gruen@drfz.de

Andreas Grützkau, German Arthritis Research Center, Schumannstrasse 20/21, 10117 Berlin, Germany; e-mail: gruetzkau@drfz.de

Thomas Häupl, Department of Rheumatology and Clinical Immunology, Charité University Hospital, Tucholskystrasse 2, 10117 Berlin, Germany; e-mail: thomas.haeupl@charite.de

Hiroshi Hata, Department of Experimental Pathology, Institute for Frontier Medical Sciences, Kyoto University, 53 Shogoin Kawahara-cho, Sakyo-ku, Kyoto 606-8507, Japan

Keiji Hirota, Department of Experimental Pathology, Institute for Frontier Medical Sciences, Kyoto University, 53 Shogoin Kawahara-cho, Sakyo-ku, Kyoto 606-8507, Japan

Rikard Holmdahl, Medical Inflammation Research, BMC I11, Lund University, S-22184 Lund, Sweden; e-mail: Rikard.Holmdahl@med.lu.se

Sally L. John, Arthritis Research Campaign Epidemiology Unit, University of Manchester, Manchester, M13 9PT, UK

Timothy D. Kayes, Department of Pathology, University of Tennessee Health Science Center, Memphis, TN 38163, USA; e-mail: tkayes@utmem.edu

Sergey V. Kozyrev, Department of Genetics and Pathology, Rudbeck Laboratory, Uppsala University, Dag Hammarskjölds väg 20, 751 85 Uppsala, Sweden

Ulf D. Landegren, Department of Genetics and Pathology/Molecular Medicine, Rudbeck Laboratory, Dag Hammarskjoldvag 20, 75185 Uppsala, Sweden; e-mail: ulf.landegren@genpat.uu.se

Kary A. Latham, Department of Medicine, University of Tennessee Health Science Center, Memphis, TN 38163, USA; e-mail: klatham@utmem.edu

Takashi Nomura, Department of Experimental Pathology, Institute for Frontier Medical Sciences, Kyoto University, 53 Shogoin Kawahara-cho, Sakyo-ku, Kyoto 606-8507, Japan

Peter Olofsson, Arexis AB, Arvid Wallgrens Backe 20, 41346 Göteborg, Sweden; e-mail: peter.olofsson@arexis.com

Robert M. Plenge, Broad Institute of MIT and Harvard, Division of Rheumatology, Allergy and Immunology, Brigham and Women's Hospital, Cambridge, MA, USA

Zhaohui Qian, Department of Medicine, University of Tennessee Health Science Center, Memphis, TN 38163, USA; e-mail: zqian@utmem.edu

Andreas Radbruch, German Arthritis Research Center, Schumannstrasse 20/21, 10117 Berlin, Germany; e-mail: radbruch@drfz.de

Edward F. Rosloniec, Departments of Medicine and Pathology, University of Tennessee Health Science Center, Memphis, TN 38163, and Veterans Affairs Medical Center, Memphis, TN 38104, USA; e-mail: erosloniec@utmem.edu

Martin Rudwaleit, Rheumatology, Department of Medicine , Charitè, Campus Benjamin Franklin, Hindenburgdamm 30, 12200 Berlin, Germany

Noriko Sakaguchi, Department of Experimental Pathology, Institute for Frontier Medical Sciences, Kyoto University, 53 Shogoin Kawahara-cho, Sakyo-ku, Kyoto 606-8507, Japan; e-mail: norikos@frontier.kyoto-u.ac.jp

Shimon Sakaguchi, Department of Experimental Pathology, Institute for Frontier Medical Sciences, Kyoto University, 53 Shogoin Kawahara-cho, Sakyo-ku, Kyoto 606-8507, Japan; e-mail: shimon@frontier.kyoto-u.ac.jp

Joachim Sieper, Rheumatology, Department of Medicine , Charitè, Campus Benjamin Franklin, Hindenburgdamm 30, 12200 Berlin, Germany; e-mail: joachim.sieper@charite.de

Takeshi Takahashi, Department of Experimental Pathology, Institute for Frontier Medical Sciences, Kyoto University, 53 Shogoin Kawahara-cho, Sakyo-ku, Kyoto 606-8507, Japan

Satoshi Tanaka, Department of Experimental Pathology, Institute for Frontier Medical Sciences, Kyoto University, 53 Shogoin Kawahara-cho, Sakyo-ku, Kyoto 606-8507, Japan

Tineke C.T.M. van der Pouw Kraan, Department of Molecular Cell Biology and Immunology, C270, VU Medical Centre, P.O. Box 7057, 1007 MB Amsterdam, The Netherlands

Cornelis L. Verweij, Department of Molecular Cell Biology and Immunology, C270, VU Medical Centre, P.O. Box 7057, 1007 MB Amsterdam, The Netherlands; e-mail: c.verweij@vumc.nl

Jane Worthington, Arthritis Research Campaign Epidemiology Unit, University of Manchester, Manchester, M13 9PT, UK; e-mail: Jane.Worthington@man.ac.uk

Ryo Yamada, Laboratory for Rheumatic Diseases, SNP Research Center (E508), RIKEN, 1-7-22 Suehiro-cho, Tsurumi-ku, Yokohama, Kanagawa-ken 230-0045, Japan; e-mail: ryamada@src.riken.go.jp

Kazuhiko Yamamoto, Laboratory for Rheumatic Diseases, SNP Research Center (E508), RIKEN, 1-7-22 Suehiro-cho, Tsurumi-ku, Yokohama, Kanagawa-ken 230-0045, Japan; and Department of Allergy and Rheumatology, Graduate School of Medicine, University of Tokyo, Tokyo 113-8655, Japan

Hiroyuki Yoshitomi, Department of Experimental Pathology, Institute for Frontier Medical Sciences, Kyoto University, 53 Shogoin Kawahara-cho, Sakyo-ku, Kyoto 606-8507, Japan

A. Introduction

Genetics of joint inflammation – problems and possibilities

Rikard Holmdahl

Medical Inflammation Research, BMC I11, Lund University, S-22184 Lund, Sweden

The search for the heredity of rheumatic diseases is not a new issue for medical science or for millions of people that are affected or have relatives that are affected. The presence of rheumatic diseases is impossible to avoid; they are easy to feel, to see and to diagnose. In addition, we nowadays have large resources of medical records and we know the genetic sequence of humans. So why have we not yet found the genetic basis of common diseases like rheumatic diseases? And, if we find the causative genes and the environmental factors, will we then be able to prevent these diseases? With these scientific tools and resources it should be appropriate today to aim for a zero tolerance for these diseases. Although, we have had only limited success so far we can now better identify the problems and see the possibilities and it should be within reach to completely eliminate many of these diseases.

This book is intended to update the state of the present knowledge of the genetic cause of rheumatic diseases. Maybe even more importantly we will describe some of the forthcoming tools, which will be used to build the bridge over the gap to fulfil the zero tolerance against rheumatic disease.

Rheumatic disease is indeed a heterogeneous group of diseases and in the book we will mainly discuss the rheumatic inflammatory diseases like rheumatoid arthritis (RA). Rheumatoid arthritis involves an inflammatory destructive attack on peripheral joints and is mainly classified using criteria reflecting signs and symptoms resulting in this attack. However, the disease starts years before this happens and the underlying factors triggering the disease or maintaining its chronicity are largely unknown. These disease pathways are likely to be quite diverse and could involve autoreactive T and B lymphocytes, activated macrophages, transformed fibroblasts and pathogenic production of cytokines. Many of these pathways are also likely to be shared with other inflammatory diseases, like ankylosing spondylitis, systemic lupus erythematosus (SLE) and type I diabetes.

To dissect the genetic causes and the disease pathways it is of critical importance to clinically define the specific subforms of these diseases, and to understand their occurrence in different populations.

The Hereditary Basis of Rheumatic Diseases, edited by Rikard Holmdahl
© 2006 Birkhäuser Verlag Basel/Switzerland

In the first chapter Jane Worthington and co-authors review the epidemiology and genetics of rheumatoid arthritis (RA), discussing the different methods to closer define the genetic contribution. As they point out there are limitations in these approaches reflecting the genetic and population complexity as well as the heterogeneity on how the disease and disease pathways are defined. Cor Verweij and co-authors continue describing how it is possible to better define the disease using gene expression analysis in which they provide evidence for subdivision of the disease and ways to define the different disease pathways – information that will be of clear importance in the further work to define the involved genes. Clearly RA is a polygenic disease (caused by many genes) and is complex (interacting with environments and genes) and has proven to be a very difficult nut to crack for geneticists. There are a few indications of some of the genes associated with various rheumatic diseases and the scientists that identified the first genes describe their findings in the forthcoming chapters. Ryo Yamada and Kazuhiko Yamamoto describe the identification of the PADI4 and the SLC22A4/A5 genes as being associated with RA in the Japanese population. The PADI4 gene is of particular importance as this codes for the enzyme critical for formation of citrullinated amino acids, recently shown to be a common autoimmune target in RA. The function of the SLC22A4/A5 gene is still unclear but interesting it interacts with a gene coding for the transcription factor *RUNX* that may also be of importance in SLE, as is discussed by Marta Alarcon-Riquelme and Sergey Kozyrev. The latter authors also describe the importance of the *PDCD1* gene in SLE and which may also play a role in RA, demonstrating shared pathways in the different rheumatic diseases. Deletion of the *PDCD1* gene has previously been shown to cause lupus in mice, which opens the possibility of closer studying of the pathogenic mechanisms. Recently, a gene coding for an intracellular tyrosine phosphatase, PTPN22, was found by Peter Gregersen and co-workers to be associated not only with SLE but also with RA, again demonstrating the sharing of pathways between inflammatory diseases.

A demonstration of the difficulties in identifying and proving the role of specific genes, and thereby opening them for studies of their role in the pathogenesis, is the major histocompatibility region (MHC). Association with MHC occurs in basically all inflammatory rheumatic diseases and has been demonstrated to be possibly the strongest association than any other gene region. However, in no cases so far has the disease associated gene in the MHC region been conclusively identified, although there are strong circumstantial evidence for MHC Class II genes in RA, MHC Class I genes in psoriasis arthritis and the Class I gene HLA B27 in ankylosing spondylitis. Of these examples the strongest evidence for association is the B27 gene, yet we do not know its precise role. This is addressed by Joachim Sieper and Martin Rudwaleit who discuss ways to understand the function of B27 and of ankylosing spondylitis. This demonstrates that even with strong evidence for the involved genes we need tools to understand their function as well as for developing preventive and therapeutic therapies based on these findings.

Possibly the greatest success of medical science during the last one or two decades has been the identification of causative genes in monogenic diseases. There are many such diseases but each of them affects a very limited number of individuals in the population. This has led, at least in some cases, to the possibility for treatment and prevention. However, maybe the most important benefit from these studies have been an increased understanding of basic biology and disease pathways, which helps us to understand the more common complex diseases. The difficulties in making similar progress in complex disease related to the fact that each contributing gene is not fully penetrant and that the genetic contribution operates in patterns of genes rather than by single genes. A way to overcome these difficulties will be to increase statistical power, i.e., to include thousands or millions of individuals in the analysis and to compare their genomes with their diseases. The technologies for such approaches are emerging but there are also ethical, logistic and analytic problems. A simplifying factor is that each individual of the human population of today consists of a mosaic of genetic fragments derived from the bottleneck of human speciation, occurring some 150–200,000 years ago. Keeping track of these genetic fragments, or haplotypes, could open the possibility to analyse which set of haplotypes are associated with specified diseases. The technology for such analysis might seem to be science fiction but is in fact already present as emerging tools, which is described by Ulf Landegren who is inventor of many of such tools and who describes the possibilities to use them in the future. Although genetics seem to be complex the advantage is that it consists of definite letter codes, just as computers, which simplifies the analysis. Obviously, from our limited success so far in cracking the genetic codes for complex diseases like rheumatic disease, it still contains considerable complexity. However, the complexity increases one step further when we analyse the expression of these genes. The expression of the genes is clearly a level between the genes and disease expression and methods to describe gene expression and relating this to diseases are also recently and rapidly developing technologies. This will clearly be useful for both diagnosis and understanding of the diseases but are still, as the genomic tools, difficult to analyse as is discussed by Thomas Häupl and co-authors.

Another, but classical, way to handle complex problems in medical science is the use of experimental animal models. Animal models, from worms to chimpanzees, have proven very relevant in studies of critical biological pathways, which is best exemplified by the discovery of cellular apoptosis in humans through studies on the microscopic worm *Caenorhabditis elegans*. Rodents have proven to be a very useful compromise between relevance and possibilities for studies of human diseases and much of the fundament of medical knowledge is today based on inbred mouse strains. Also, models for rheumatoid diseases have proven to be useful and mouse models are today required by FDA (Food and Drug Administration) in the US for the development of new therapies for RA. There has been a rapid development in developing technologies in using such animal models. One of these tools is the possibility to genetically manipulate the animal genome, allowing the possibility to make controlled genetic changes in the genome and study its effect on the complex *in vivo* situation. One example on how this can be used is discussed by Kary Latham, Edward Rosloniec and colleagues who describe how mice can be 'humanised' by inserting relevant human MHC Class II genes and study their impact on models for rheumatoid arthritis. This enables possibilities to confirm the role of human genes and to study their function and therapeutic possibilities *in vivo*. These studies are based on the strikingly similar results using the corresponding mouse genes and will be a powerful tool for further studies. There is, however, also a raising awareness that genetic manipulation of mice may also introduce abnormalities that divert the results from reality. Introduction of deletions or the introduction of foreign genes may not cooperate well with the rest of the mouse genome. An alternative look at this problem would be to use the mice as we use the humans, to search for the naturally polymorphic DNA sequences that cause disease, with the assumption that mice and humans are close enough to share some of the major disease pathways. Shimon Sakaguchi and co-authors give one example in which they describe the search for genes causing a spontaneous development of arthritis in mice. The major causative gene was identified to be ZAP70, in which the arthritis-prone mice had an amino acid replacement mutation. This clearly points out the threshold for T cell receptor signalling as being critical for development of arthritis and the model is now open for further analysis of the pathways as well as translating the results into the corresponding pathway affecting human arthritis. Another way to identify relevant genes is described by Peter Olofsson who describe the use of crosses of inbred rat strains that are susceptible *versus* resistant to arthritis. Using such crosses it is possible to make linkage analysis with a power thousand-fold larger than the human studies, since the strains are inbred, the disease more precisely characterised and the environment better controlled. They could identify one of the genes of importance for development of arthritis severity, Ncf1, which controls a pathway regulating oxidative burst. In fact, lower oxidative burst led to increased arthritis severity. He also shows that this finding could be utilised to develop a new therapeutic strategy that could have potential for treatment of human disease. Thus, identification of the genes in the animal models may not only enable identification of genes in complex disease, but also provides a tool to shorten the time to develop the findings into therapy for human disease. Obviously, it will in most cases not be the genetic polymorphism *per se* that is the critical target for prevention or therapy but rather the critical pathways leading to disease.

B. Genetic studies on rheumatic diseases

The epidemiology of rheumatoid arthritis and the use of linkage and association studies to identify disease genes

Jane Worthington, Anne Barton and Sally L. John

Arthritis Research Campaign Epidemiology Unit, Division of Epidemiology and Health Sciences, University of Manchester, Oxford Road, Manchester, M13 9PT, UK

Epidemiology of rheumatoid arthritis

Rheumatoid arthritis (RA), the most common form of chronic inflammatory polyarthritis, represents a significant health burden in the developed world. The damage and deformity of the synovial joints characteristic of RA most commonly develops in the sixth decade but can occur at any age and will usually require treatments and interventions for the rest of an individual's life [1]. It is diagnosed and distinguished from other arthritic diseases on the basis of criteria defined in 1987 [2]. The criteria are based on factors such as the presence of the autoantibody, rheumatoid factor, erosion of joints visible upon x-ray, stiffness, swelling and symmetry of affected joints. Interestingly, with improved treatments, many patients will display some of these features only transiently and would strictly speaking only satisfy criteria if they were applied cumulatively, as has been proposed when the criteria are used in epidemiological studies [3] (Tab. 1).

Rheumatoid arthritis has a worldwide prevalence of approximately 1% (prevalence is the number of cases occurring in a population at a given time) and is consistently observed to affect women 2–3 times more frequently than men. The occurrence of RA is not, however, the same throughout the world [4]. Prevalence rates are low in the less developed and rural parts of the world and it has been suggested that RA is a modern disease, its appearance seeming to coincide with industrialisation or urbanisation. A study in South Africa found a low frequency of RA among Bhantu-speaking people in their traditional rural environment but higher rates in the same ethnic group living in the modern urban townships of Soweto, similar in fact to Caucasians living in nearby Johannesburg [5, 6]. This apparent influence of urbanisation was not however observed in a study comparing rural Chinese with those living in the highly industrialised society of Hong Kong. The frequency of RA was low in both the Chinese populations studied. Other factors such as diet and a lower or different genetic susceptibility, may explain these apparently contradictory findings.

In common with other autoimmune or chronic inflammatory conditions, most notably multiple sclerosis, there appears to be a latitude related gradient for the

The Hereditary Basis of Rheumatic Diseases, edited by Rikard Holmdahl
© 2006 Birkhäuser Verlag Basel/Switzerland

Table 1 - Modification to 1987 ARA criteria to improve ascertainment of inactive cases. (With permission by the Journal of Rheumatology, from [3])

Criterion	Current	Ever
1. Morning stiffness in and around the around the joints lasting at least 1 h before maximal improvement	Reported in the 6 weeks prior to interview	At any time in the disease course
2. Arthritis of three or more joint areas	With soft tissue swelling or fluid at current examination	With swelling at current examination or deformity and a documented history of swelling
3. Arthritis of hand joints: Involvement of at least one area in a wrist MCP or PIP joint	With soft tissue swelling or fluid at current examination	With swelling at current examination or deformity and a documented history of swelling
4. Symmetrical arthritis: Simultaneous involvement of the same joint areas on both sides of the body	With soft tissue swelling or fluid at current examination	With swelling at current examination or deformity and a documented history of swelling
5. Rheumatoid nodules: Subcutaneous nodules	Present at current examination	Present at current examination or documented in the past
6. Serum rheumatoid factor	Detected at current examination	Present at current examination or documented to have been positive in the past by any assay method
7. Radiographic changes	Erosive changes typical of RA on postero-anterior hand and wrist radiographs	Erosive changes typical of RA on postero-anterior hand and wrist radiographs

Individuals satisfying four or more criteria classified as having RA prevalence of RA with rates being highest at more northerly latitudes. In Finland, France and Italy rates for men and women are; 0.6% and 1%, 0.32% and 0.86%, 0.13% and 0.51% respectively [7]. Many factors, both genetic and environmental, such as such as climate, ultraviolet (UV) exposure, diet could contribute to this observation.

There is some evidence that prevalence rates, particularly in women, are declining but monitoring of patterns of incidence of disease (i.e., the number of new cases in a given time period) can be more sensitive to changing patterns and may give clues as to the aetiology of disease. A number of studies have shown a fall in the incidence of RA over the last four decades [8–10]. In the US (Rochester, Minnesota) the incidence rate in women decreased from 83/100,000 in 1955–1964 to 40/100,000 in 1985–1994 [8]. One explanation put forward for this fairly rapid change is the protective effect of the oral contraceptive pill (OCP). This hypothesis provides an appealing explanation for the gender differences observed and data collected over the next couple of decades should be revealing, as one would expect that, as OCP use reaches saturation, incidence rates should stabilise.

In contrast to the idea that RA is a modern disease is the observation of skeletons found in North America dating back several thousand years, showing evidence of RA [11]. This finding coupled with the fact that some of the highest prevalence rates for RA are observed in Native American peoples has led to the suggestion that RA may have been introduced to The Old World by explorers returning to Europe from the New World. The incidence and prevalence of RA in Pima Native Americans has fallen substantially in recent years and it will be interesting to observe future rates in developing populations in which RA seems to have arrived more recently than in Europe [12]. A number of studies also suggest that RA is becoming less severe although teasing this out in the context of ever improving therapies and patient management could prove difficult [13].

Genes *versus* environment

It is now well established that RA is a complex disease for which susceptibility and severity is likely to be influenced by combinations of both genetic and environmental factors. The environmental factors implicated in RA include obesity, diet, smoking and a number of infectious agents including parvovirus, proteus and various retroviruses [13, 14]. Interestingly the pattern of increased risk associated with the non-infectious agents is common to various cancers and other chronic inflammatory conditions such as cardiovascular disease.

Most studies investigating risk factors have however been carried out with cross-sectional case identification and analysis based on retrospective data, making for significant difficulties and possible bias. In contrast The Norfolk Arthritis Register (NOAR) in the UK is a prospective study of RA initiated in 1989 with recruitment through general practitioners [1]. This dataset confirmed the association with obesity and smoking but there was only weak evidence of parvovirus infection [15]. Interestingly there was a higher than expected frequency of tetanus immunisation in the 6 weeks prior to onset and an association with prior blood transfusion again suggesting that various challenges to the immune system could be precipitating events. The geographical overlap of NOAR and part of the European Prospective Investigation of Cancer (EPIC) provided a unique opportunity to study diet, based on a 7-day food record collected prior to the onset of arthritis. Cases were found to have a low vitamin C and high red meat intake compared to controls matched for age, gender and time of recruitment to EPIC [16, 17].

While some of the population differences described above might appear to be most readily explained by environmental exposures, underlying genetic factors clearly have a role. This is best illustrated by the *HLA-DRB1* gene locus, the major genetic factor associated with RA, to date. The definition of the shared epitope (SE), the sequence of amino acids common to *HLA-DRB1* alleles associated with RA, arose out of HLA association studies carried out in different populations and it is widely reported that SE positive *HLA-DRB1* alleles are at a low frequency in populations in which RA is relatively rare (for full review see Chapter 2). The relative importance of genetic and environmental factors and the contribution of individual loci can be investigated using twin and family based studies.

Twin and family studies

An increased occurrence of disease in first-degree relatives of an affected case is often cited as evidence for a genetic component to susceptibility although it might also be in part attributed to a shared environment. Many studies have demonstrated an increased risk to siblings of RA probands when compared to a control panel, with relative risk ranging from 1.6–15.8 in studies in which the cases were derived from hospital series (reviewed in [18]). The lowest sibling risk recorded (1.2) was from the only population-based study [19]. Meta-analysis of this data suggests an overall increased risk of two [19]. The sibling risk ratio (λ_s) [20] is derived by dividing the sibling risk by population prevalence and is the value most frequently used when comparing the size of the genetic component of susceptibility across diseases. For RA λ_s is said to be between 2 and 15 depending on the values taken for both sibling risk and population prevalence. These values should ideally be derived from the same population and in only study to date in which this has been done λ_s was found to be 10 [21, 22].

Studies in twins have also been used to estimate the relative proportions of genetic and environmental components of disease susceptibility. The assumption is that both monozygotic (MZ) and (same-sexed) dizygotic (DZ) twin pairs will share environmental exposure to the same extent thus any excess concordance in MZ pairs would be due to genetic factors. The upper limit of concordance in MZ twin pairs allows estimates of the maximum size of the genetic component. There are however a number of potential biases to twin studies. Many twin studies are cross sectional and therefore tend to underestimate concordance, particularly for diseases such as RA with a late age at onset. It is now accepted that early twin studies overestimated concordance rates at 30% MZ, 5% DZ [23] and more recent studies have reported lower levels of 15.4% MZ, 3.6% DZ [24].

Identification of genes for complex diseases

The search for genetic variants determining disease susceptibility is based on the use of two different scientific strategies: linkage and association studies. For complex diseases, extended multi-case pedigrees are rarely available and the most commonly used linkage analysis is based on affected sibling pair families. The analysis of hundreds of highly informative microsatellite markers across the genome in large numbers of families, allows the identification of regions of the genome shared identical by descent more frequently by the affected siblings than expected by chance and, thus, likely to contain disease genes. By contrast, association studies rely on comparison of polymorphic markers in cohorts of cases and matched unaffected controls, a difference in allele or genotype frequency providing evidence that the polymorphism itself or one in linkage disequilibrium with it, is associated with disease. There are pros and cons to both approaches and the tide of opinion as to the merits of either is constantly influenced by developments in bioinformatics, genotyping technologies and genetic analysis. In the following sections we discuss the advantages of both linkage and association analysis at the current time and how future developments might further influence the use of both methodologies.

Linkage analysis

Linkage analysis methods have traditionally been applied to the search for novel susceptibility loci across the entire genome rather than to individual candidate genes. Collections of a few hundred affected sibling pairs (ASPs) genotyped for highly polymorphic markers spaced at regular intervals (10 cM) across the genome has been the bread and butter of researchers attempting to unravel the genetics of complex disease. The aim of these scans is to identify regions of the genome, typically several centi-Morgans (1 cM is approximately equivalent to 1 Mb in physical distance), that are inherited by affected siblings more often than one would expect under random Mendelian inheritance [25]. The allure of such whole genome scans

(WGS) has been the promise of the discovery of novel genes that would not have been considered in any candidate gene study undertaken in an association analysis. There are, however, other challenges posed by linkage studies that need to be considered, most of which can be illustrated with reference to published linkage studies in RA.

Five WGS using evenly spaced microsatellite markers have been published in RA to date from four different populations. One can distil two main points from these studies, firstly that a locus containing HLA is convincingly detected in most scans and, secondly, no other locus stands out across studies. This provides good evidence that the effect sizes of non-HLA-RA susceptibility genes are much smaller than HLA. It has been estimated that HLA accounts for 50% of the genetic component of RA, it is therefore likely that many genes make up the remaining 50%. It is difficult to estimate how many other loci there may be, however, Jawaheer et al. added up the effect size of each of the loci detected in a WGS of 276 North American RA consortium (NARAC) families and identified five loci on chromosomes 1, 12, 16, 17 and 4, that when combined accounted for the majority of the familial clustering observed for RA, suggesting that this would be the minimum number of loci expected [26].

Linkage studies using ASP methods require much larger sample sizes than have been analyzed so far, in order to detect genes with modest genetic effects (genotype relative risk [GRR] <2) [27]. It is the collection of sufficient numbers of suitable families that impedes the success of linkage studies most and this problem is to a large extent intractable. For this reason meta-analysis methods are appealing as it substantially increases the power to detect loci that are not influenced by genetic heterogeneity. However, when one looks at these scans in more detail, there is very little overlap in terms of study design and analysis methodology. The smallest was a Japanese study that consisted of 41 families and performed a single point analysis [28]. The European consortium of RA families (ECRAF) study employed a multi-point analysis of markers at a 10 cM spacing on 97 families, followed by additional analysis using a denser marker set [29, 30]. The US (NARAC) and UK, studies utilised 677 (512 families) and 469 (377 families) ASPs, respectively, and took a two-stage approach, splitting the cohorts roughly in half and performing a WGS in an initial cohort and then attempting to replicate the results in an independent cohort across the whole genome or at loci showing an increase in allele sharing [31, 32]. Significant linkage ($p < 3 \times 10^{-5}$) to the HLA region on chromosome 6p was observed in all but the Japanese study, which did nonetheless detect linkage at a lower significance threshold [33]. Nominal evidence for linkage ($p < 0.05$) to regions on chromosomes 1q and 14q was also observed in all of the studies of Caucasians (Fig. 1). In all four populations different analytical tools were employed to identify areas of the genome that are linked to disease. It is well known that when several analytical methods are applied to the same data set, results can be quite different [34, 35]. It is issues such as these that confound the interpretation of results across

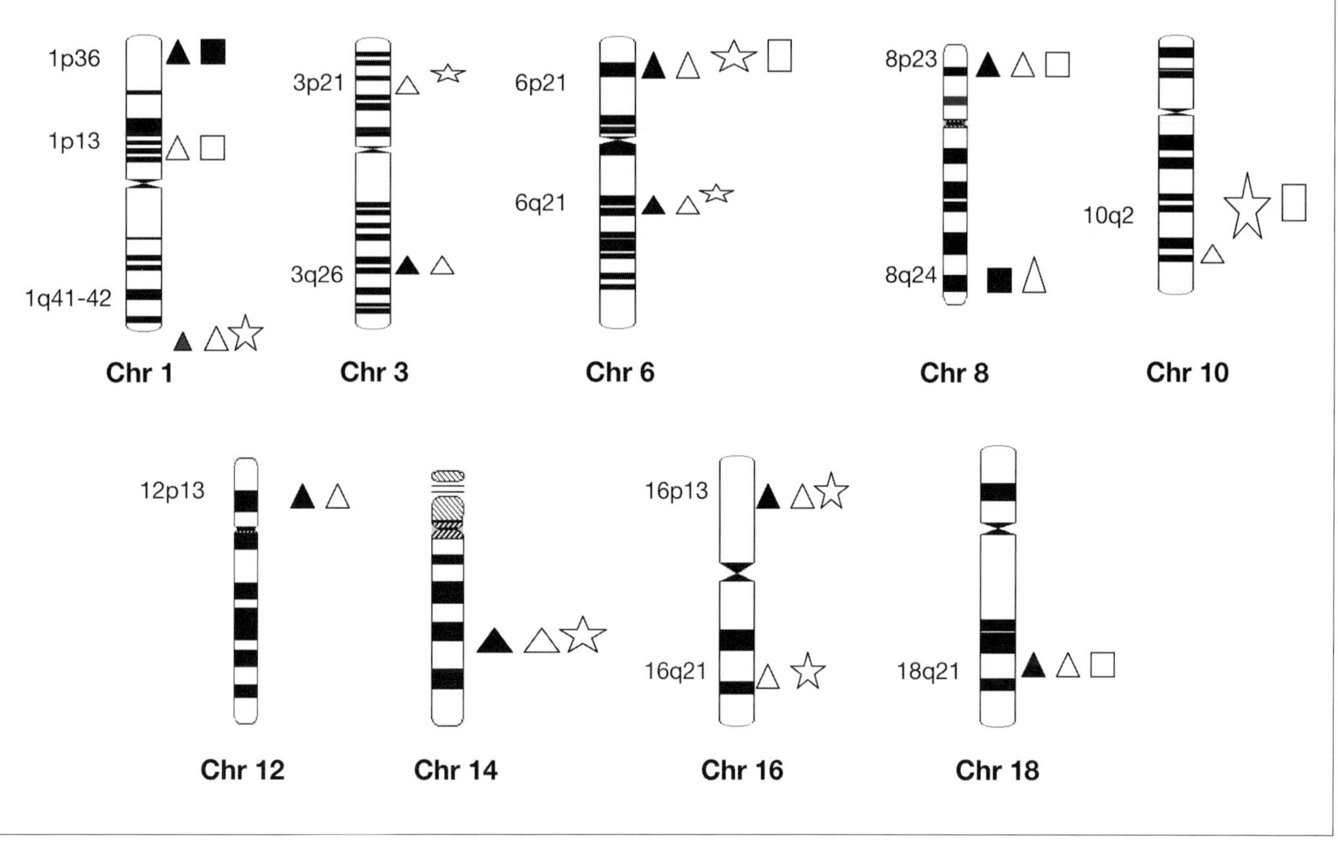

Figure 1

Diagrammatic representation of regions of overlap in WGS of European (▲), Japanese (■), US (△ = cohort 1, □ = cohort 2) and UK (☆) populations studies, making it particularly difficult to determine the extent of genetic heterogeneity between populations.

One approach to overcome the lack of statistical power is to pool data from the various studies that have already been performed [36]. Recently a Genome Screen Meta-Analysis (GSMA) of four of the above studies reported a novel locus on chromosome 16 [37]. This method of meta-analysis divides the genome into bins and ranks the significance of each bin within each study, totalling the ranks for each bin to give a global score. The GSMA approach has appeal; it allows results from multipoint and single-point analysis, parametric and non-parametric, as well as results from different analysis methods (such as LOD scores *versus* NPL scores) and different marker sets to be combined [38]. However, as the authors indicate, the optimal approach is pooled analysis of the original genotype data. These efforts are currently underway in linkage analysis and will allow us to prioritise regions of linkage for further investigation.

Recent advances in technology and statistical methodology have allowed researchers to address some of the limitations of affected sibling pair methods. Multi-case RA families in common with other family collections of late age at onset disease tend to have limited parental information. As a result the actual amount of information about inheritance (information content) that can be extracted is well below the predicted theoretical maximum, given the marker heterozygosity. The development of a micro-array allowing simultaneous genotyping of > 10,000 single nucleotide polymorphisms (SNP) has allowed family collections to be re-analyzed using a very dense SNP map [39]. This dramatically increases the information content especially in families with a high proportion of missing parents, leading to an increase in LOD scores (the statistical measure of linkage) [40, 41]. A secondary benefit is the narrower linkage interval that is often observed resulting in a smaller region of the genome to be followed up in fine mapping studies [41].

Although there are only rare examples to date of WGSs carried out in RA leading to the positional cloning of candidate genes (e.g., the *PADI4* gene), the information gained as a result of WGS will still make a significant contribution to the prioritisation of regions of the genome most likely to harbour susceptibility genes and the interpretation of results from association studies [42]. As WGS by default have information about inheritance across the entire genome, one is able to employ analytical methods that detect interactions between loci. Recently we have re-analysed the UK genome-wide linkage data, conditional on a sibling's allele sharing status at *HLADRB1* and two loci on chromosome 6q and 16p showed substantially increased evidence of linkage suggesting that there may an epistatic interaction between genes at these loci and HLA (unpublished data). The inherent low power of linkage analysis to detect common loci conferring low relative risks suggests that any loci detected by linkage will be more easily detectable by association and prioritising these regions may prove to be a valuable strategy.

Association studies

As outlined above, linkage studies have been less successful than originally hoped in identifying the genes underlying complex diseases. Reasons for this include the low heritability of most common diseases and the heterogeneity of phenotypes investigated. This is particularly true of many rheumatic diseases, which show wide variation in their clinical presentation, progression, severity and associated autoantibody profiles. However, the major reason for the lack of success of linkage-based approaches is because they are inherently underpowered to detect genes with modest effect sizes where allele frequencies are common. It is likely that most effect sizes important in common disease will be in the order of < 1.5 and that disease associated alleles will be common (> 10%). By way of example, both linkage and association studies have consistently identified the chromosome 6p region as a susceptibility locus for RA. The *HLA DRB1* gene is estimated to have an effect size of ~3 and frequency of shared epitope carriage is common. However, where disease alleles are rare but have large effect sizes (e.g., *CARD15* gene variants causing susceptibility to Crohn's disease) linkage studies are more likely to identify the locus than association studies. By contrast, for the situations expected for most complex diseases where effect sizes are small but disease causal alleles occur at a reasonable frequency in the population, association will have much greater power to detect a disease locus than linkage studies (Fig. 2). Secondly, because association relies on the existence of linkage disequilibrium (LD) between the disease and marker allele and LD only extends for short distances, association based methods have greater power to localise a disease gene. Finally, studies on the association between genetic variants and disease can suggest pathogenic mechanisms and have the potential for direct clinical application by providing markers of risk, diagnosis, prognosis, and, possibly, therapeutic targets [43].

A number of genes underlying common complex rheumatic disorders have been identified (Tab. 2). Those studies all adopted a candidate gene approach in which genes or regions are investigated based on prior knowledge of biological pathways, information gleaned from animal models or other similar diseases or because they map within a region of linkage. Most association studies have investigated single nucleotide polymorphism (SNP) markers because these are abundant in the genome, amenable to high throughput genotyping and may affect gene function. In order to detect association with a disease gene, the markers tested must either be causal or strongly correlated with the causal variant. It is estimated that up to 80% of the genome falls into segments of high LD within which variants are strongly correlated with each other and most chromosomes carry one of only a few common haplotypes. It is, therefore, inefficient to genotype all the SNPs within a gene. Cost efficiency is gained by genotyping a subset of markers, tagging SNPs, which capture most of the allelic variation in a region [44]. The availability of information regarding LD patterns across the genome with which to select tagging SNPs has been facil-

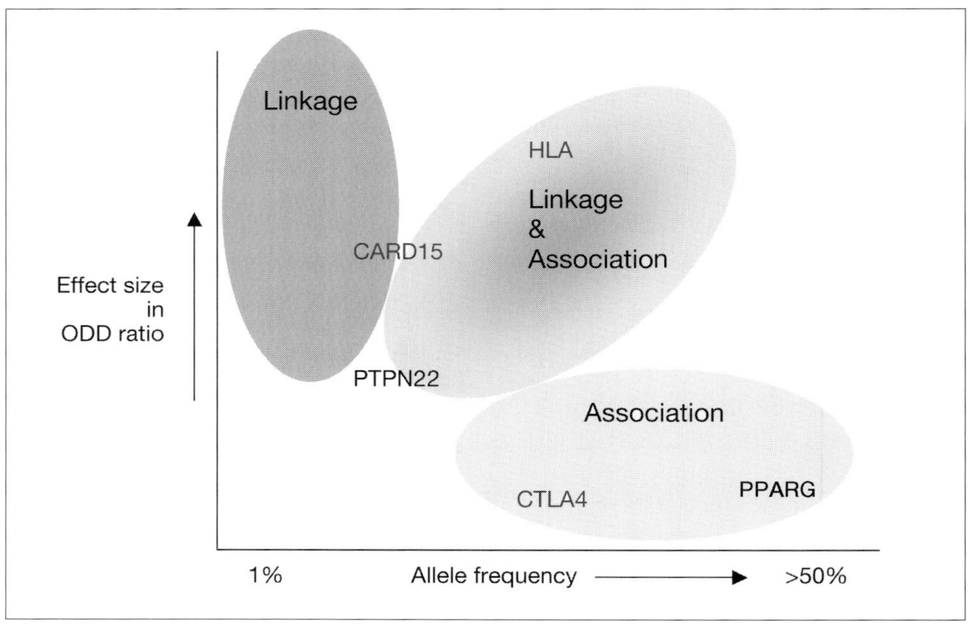

Figure 2
Linkage and association methods may be more suited to detecting genes depending upon the allele frequency and effect size, assuming typical sample sizes of several hundred families or case-control pairs. Using currently available methods, it is unlikely that rare alleles (<3%) conferring small risks (<1.2) will be detected in any study.

itated by initiatives such as the HapMap project, which has created a genome-wide catalogue of common haplotype blocks in multiple human populations [45].

An alternative approach would be only to genotype those SNPs encoding missense mutations, the argument being that these are most likely to be disease causal and the number of SNPs requiring genotyping would be markedly reduced [46]. This approach has been successful, for example, in identifying the *PTPN22* gene, which has been associated with a number of autoimmune diseases (Chapter 8). However, for diseases of late age at onset including many common diseases, non-coding regulatory variants are likely to play an important role as they are less likely to have dramatic effects on disease risk and are less likely to be subject to strong negative selection. Examples are emerging in the literature to support this viewpoint. For example, regulatory variants in the *CTLA-4* gene have been associated with susceptibility to autoimmune thyroid disease and type 1 diabetes [47].

Many gene-disease association studies have been performed, a large number of positive associations have been reported but few have been widely replicated and

Table 2 - Genes widely replicated with functional support in rheumatology

Gene	Disease	Population	Polymorphism	Refs
PTPN22	RA, SLE, T1D, AIT, JIA	Europeans/US	missense mutation	[64–69]
PDCD1	SLE	Europeans/US	intron 1	[70]
PADI4	RA	Japanese/Korean	haplotype	[42]
FCRL3	RA, SLE, AITD	Japanese	promoter variant	[71]
MIF	JIA	European	promoter haplotype	[72, 73]

identified disease causal mutations. Not excluding the possibility that results have arisen by chance is probably the major reason why many apparent associations are not replicated [48]. This arises because information about SNPs is increasingly available, genotyping costs continue to fall, more tests are performed and, hence, more are likely to show association by chance alone (false positive) if significance thresholds are not appropriately adjusted. Applying a Bonferroni correction for SNPs that are tightly linked is overly conservative and permutation testing has been proposed as an alternative to empirically test the probability of having observed an association by chance [49]. Thresholds for declaring significant association also need to be reviewed in light of the increasing number of hypotheses being tested in candidate gene studies. Suggestions have been made to adjust the p-value for declaration of statistical significance from 5×10^{-2} to 5×10^{-5} in candidate gene association studies [48, 49]. This will require at least a three-fold increase in sample size to retain the same power [48]. Increasing the prior probability of detecting association should also reduce the false positive rates [50]. This can be achieved by selecting genes for investigation, which either map to regions of linkage or where there is evidence for association in a different population.

There is a wealth of data available from published whole genome linkage scans for a number of rheumatic diseases including osteoarthritis, systemic lupus erythematosus and, as outlined above, RA. Furthermore, when selecting genes for disease association studies, some consideration should be given to the biological plausibility. For example, the gene should be expressed in a relevant tissue or be involved in the disease pathway. Large well-powered studies are required in order to robustly assess the significance of results using stringent criteria. A small sample size (hence, low power to detect anything other than major effects) in a first reported association has previously been shown to predict inconsistent replication in subsequent studies [51]. Indeed only two of seven studies reporting positive association in which the sample size was less than 150 were subsequently replicated [52].

Replication of findings in an independent cohort provides compelling evidence that the original association was real. It should be noted that replication studies should be powered to detect a smaller effect size than that reported in the original study because of the phenomenon of 'winner's curse' whereby there is an upward bias in the effect sizes reported in original studies [53]. This has been demonstrated in an investigation of 55 meta-analyses, which showed that subsequent research suggested weaker or no association compared with strong associations suggested by first research [51].

Population stratification has been proposed as another reason why false positive associations may occur. This arises when a population studied actually comprises subpopulations that differ both in allele frequency and in the prevalence of the disease under study. With more information publicly available regarding SNP frequencies in different populations, it is increasingly recognised that differences between populations exist. Very little difference is seen between individuals of Northern European descent and white US groups but Hispanics, African descent and Asians can have very different allele frequencies [54]. This can have important consequences for association. For example, the *PTPN22* gene missense mutation has been widely associated with a variety of autoimmune diseases in US and European populations but the polymorphism does not exist in Asian populations. However, despite these concerns, several studies have now demonstrated that major population stratification is unlikely to be a problem in well-matched case control cohorts [55] and methods exist to correct for it if it is present by genotyping a limited number of markers in unlinked regions of the genome [56]. An alternative is to use family based methods such as the transmission disequilibrium test, which eliminates the possibility of stratification. However, these methods are less powerful than case control methods, are inefficient in their use of available information and are more prone to technical artefacts [49]. Furthermore, when investigating gene-environment interactions, overmatching of environmental exposures may occur using family-based methods [43]. Hence, there has been a resurgence of interest in using large, well-powered case-control study designs to investigate associations.

The power of a study to detect an association with disease depends on both the sample size studied and the frequency of the alleles that predispose to disease. If a susceptibility allele has a minor allele frequency of less than 10% in the population and a small effect size, many thousands of individuals will be required to detect association with disease. It is estimated that most effect sizes important in common disease will be in the order of < 1.5. The 'common disease common variant' hypothesis supposes that, because common diseases generally arise later in life and thus do not have a major effect on reproductive ability, the variants responsible are unlikely to be subject to negative selection and will have reached a reasonably high frequency in the population. However, even if this is true and common diseases are caused by alleles that have allele frequencies > 10%, large sample sizes are still crucial to detect modest effect sizes. In practice, it is likely that some diseases will have associated genes with low minor allele frequencies (but these may have a large effect size), while others will be due to common variants with small effect sizes [57, 58].

Once a genetic association has been robustly demonstrated, it is important to determine whether the associated variant is disease causal or highly correlated with a disease causal polymorphism. New approaches for the assessment of the functional significance of associated variants using statistical, bioinformatics and molecular biology approaches are being developed and should aid in the interpretation of associations between genotypes and disease [59].

It is clear that well-designed candidate gene association studies show great promise in contributing to the identification of the underlying genetic causes of common diseases. Indeed, there have already been a number of successes using this approach but also many reported associations that have not been replicated. A number of recommendations regarding study design and guidelines for referees in interpreting the quality of reports are now available [43, 48, 60]. In general terms, these highlight the importance of good lab practice (including use of negative controls and duplicates, blinding of lab personnel), rigorous statistical analysis (assessment of whether genotype frequencies conform to Hardy-Weinberg expectations, power and effect size considerations, correction for multiple testing), replication in independent data sets and assessment of biological plausibility.

The future

Association studies performed, to date, have been restricted to candidate gene/region studies. As the number of SNPs in the public domain increases, there is a move to perform association studies with SNPs spanning the entire genome (whole genome by association [WGA]). The main argument for this approach is that it is hypothesis free; hence, no prior knowledge of gene function is required. No truly genome-wide studies have been published yet but many are planned. The first will probably use chip-based genotyping, for example using chips containing 100,000 SNPs (www. Affymetrix.com) or bead-based methods (www.illumina.com). It is unclear how much of the genome will actually be captured but it is estimated to be in the region of 50%. Other companies are developing methods for higher throughput sequencing and it may be that whole genome sequencing will become standard in the future (e.g., www.curagen.com). Opponents of this approach point to the large number of false positives likely to be generated. This can be countered by setting stringent significance thresholds but, in turn, this reduces the power to detect modest effect sizes unless large sample sizes are used. Most argue for a multi-stage approach whereby only SNPs for which association is replicated are explored for functionality.

Although whole genome linkage analyses have had limited success, the data generated from such studies will continue to be of use in the search for important loci. While the statisticians develop strategies for the analysis of WGA data that make some correction for multiple testing without completely removing the possibility of detecting weak associations, it may be useful to focus analysis of such data on regions already known to be linked to the disease of interest. The DNA collected from families with multiple cases of RA will also prove to be a valuable resource. These cases tend to have a lower age at onset and more severe disease than their community based counterparts and are likely to have a higher 'dose' of genetic risk factors increasing the likelihood of detecting genes in these cohorts.

In considering the merits of linkage and association strategies it is clear that there is no simple answer as to which should be the method of choice. For the disease of interest it is important to take account of a number of factors, including feasibility of collecting multi-case families, the expected effect size of disease genes, the allele frequency of associated polymorphisms and the budget available. For major genes with significant effect sizes (e.g., HLA with RA) linkage analysis of small number of families or a modest case-control association study would both be successful. For disease alleles found only at low frequency in the population of interest, linkage would be powerful (e.g., the *CARD15* gene and Crohn's disease) whereas a disease allele with high frequency would rarely be detected by linkage but could be found by an appropriately powered association study (e.g., the *PPARG* gene and type 2 diabetes) [49] (Fig. 2).

Current efforts are focused mainly around identifying susceptibility loci for complex diseases in an effort to gain a better understanding of the underlying disease pathogenesis and, ideally, to develop safer, more effective treatments. In the future, however, other potentially clinically useful applications arising from the identification of disease genes include firstly, the identification of severity genes so that interventions can be better targeted. There are now a number of inception cohorts of patients with early disease being followed prospectively across Europe and this should facilitate the separation of susceptibility from severity factors. For example, patients with early inflammatory or rheumatoid arthritis are being recruited in centres in the UK, The Netherlands and Sweden [1, 61, 62]. Secondly, it is clear that many of the rheumatic diseases are clinically heterogeneous and may in fact comprise a number of distinct conditions with a similar end-phenotype, in the case of RA, for example, symmetrical synovial joint inflammation. Once genetic susceptibility factors have been identified, it may be possible to use this information to perform phenotyping studies to identify these subgroups as has been done in Crohn's disease by analysis of *CARD15* gene variants [63]. Finally, knowledge of genetic susceptibility factors may influence response to or susceptibility to side effects from particular treatments. In this respect, the setting up of Biologics Registers in many European countries is encouraging and will provide an ideal opportunity to establish whether treatment response genes exist.

Acknowledgements
The Arthritis Research Campaign and The Wellcome Trust for their support of J Worthington and A Barton, respectively.

References

1   Symmons DP, Silman AJ (2003) The Norfolk Arthritis Register (NOAR). *Clin Exp Rheumatol* 21: S94–S99

2   Arnett FC, Edworthy SM, Bloch DA, McShane DJ, Fries JF, Cooper NS, Healey LA, Kaplan SR, Liang MH, Luthra HS et al (1988) The American Rheumatism Association 1987 revised criteria for the classification of rheumatoid arthritis. *Arthritis Rheum* 31: 315–324

3   MacGregor AJ, Bamder S, Silman AJ (1994) A comparison of the performance of different methods of disease classification for rheumatoid arthritis. Results from an analysis from a nationwide twin study. *J Rheumatology* 21: 1420–1426

4   Abdel-Nasser AM, Rasker JJ, Valkenburg HA (1997) Epidemiological and clinical aspects relating to the variability of rheumatoid arthritis. *Semin Arthritis Rheum* 27: 123–140

5   Beighton P, Solomon L, Valkenburg HA (1975) Rheumatoid arthritis in a rural South African Negro population. *Ann Rheum Dis* 34: 136–141

6   Solomon L, Beighton P, Valkenburg HA, Robin G (1975) Proceedings: Rheumatoid arthritis in the South African Negro. *Ann Rheum Dis* 34: 463–464

7   Lau E, Symmons D, Bankhead C, MacGregor A, Donnan S, Silman A (1993) Low prevalence of rheumatoid arthritis in the urbanized Chinese of Hong Kong. *J Rheumatol* 20: 1133–1137

8   Doran MF, Pond GR, Crowson CS, O'Fallon WM, Gabriel SE (2002) Trends in incidence and mortality in rheumatoid arthritis in Rochester, Minnesota, over a forty-year period. *Arthritis Rheum* 46: 625–631

9   Kaipiainen-Seppanen O, Aho K, Isomaki H, Laakso M (1996) Incidence of rheumatoid arthritis in Finland during 1980-1990. *Ann Rheum Dis* 55: 608–611

10 Shichikawa K, Inoue K, Hirota S, Maeda A, Ota H, Kimura M, Ushiyama T, Tsujimoto M (1999) Changes in the incidence and prevalence of rheumatoid arthritis in Kamitonda, Wakayama, Japan, 1965–1996. *Ann Rheum Dis* 58: 751–756

11 Rothschild BM, Woods RJ, Rothschild C, Sebes JI (1992) Geographic distribution of rheumatoid arthritis in ancient North America: implications for pathogenesis. *Semin Arthritis Rheum* 22: 181–187

12 Rothschild BM (1992) Relationship of Christopher Columbus to spread of rheumatoid arthritis. *Arch Intern Med* 152: 1730

13 Silman AJ, Pearson JE (2002) Epidemiology and genetics of rheumatoid arthritis. *Arthritis Res* 4 (Suppl 3): S265–S272

14 Symmons DP, Bankhead CR, Harrison BJ, Brennan P, Barrett EM, Scott DG, Silman AJ (1997) Blood transfusion, smoking, and obesity as risk factors for the development of rheumatoid arthritis: results from a primary care-based incident case-control study in Norfolk, England. *Arthritis Rheum* 40: 1955–1961

15 Symmons D, Harrison B (2000) Early inflammatory polyarthritis: results form the Norfolk Arthritis Register with a review of the literature. I. Risk factors for the

development of inflammatory arthritis and rheumatoid arthritis. *Rheumatology (Oxford)* 39: 835–843

16 Pattison DJ, Symmons DP, Lunt M, Welch A, Luben R, Bingham SA, Khaw KT, Day NE, Silman AJ (2004) Dietary risk factors for the development of inflammatory polyarthritis: evidence for a role of high level of red meat consumption. *Arthritis Rheum* 50: 3804–3812

17 Pattison DJ, Silman AJ, Goodson NJ, Lunt M, Bunn D, Luben R, Welch A, Bingham S, Khaw KT, Day N, Symmons DP (2004) Vitamin C and the risk of developing inflammatory polyarthritis: prospective nested case-control study. *Ann Rheum Dis* 63: 843–847

18 Seldin MF, Amos CI, Ward R, Gregerson PK (1999) The genetics revolution and the assault on rheumatoid arthritis. *Arthritis Rheum* 42: 1071–1079

19 Jones MA, Silman AJ, Whiting S, Barrett EM, Symmons DP (1996) Occurrence of rheumatoid arthritis is not increased in the first degree relatives of a population based inception cohort of inflammatory polyarthritis. *Ann Rheum Dis* 55: 89–93

20 Risch N (1987) Assessing the role of HLA-linked and unlinked determinants of disease. *Am J Hum Genet* 40: 1–14

21 Ward RH, Hasstedt SJ, Clegg DO (1992) Familial aggregation of RA: genetic lliabiliy of a threshold trait. *Arthritis Rheum* 35: S127

22 Ward RH, Hasstedt SJ, Clegg DO (1992) Population prevalence of rheumatoid arthritis is lower than previuosly supposed. *Arthritis Rheum* 35: S126

23 Lawrence JS (1970) Heberden Oration, 1969. Rheumatoid arthritis--nature or nurture? *Ann Rheum Dis* 29: 357–379

24 Silman AJ, MacGregor AJ, Thomson W, Holligan S, Carthy D (1993) Twin concordance rates for rheumatoid arthritis: results from a nationwide study. *Brit J Rheum* 32: 903– 907

25 Penrose LS (1955) Evidence of heterosis in man. *Proc R Soc Lond B Biol Sci* 144: 203– 213

26 Jawaheer D, Seldin MF, Amos CI, Chen WV, Shigeta R, Monteiro J, Kern M, Criswell LA, Albani S, Nelson JL et al (2001) A genomewide screen in multiplex rheumatoid arthritis families suggests genetic overlap with other autoimmune diseases. *Am J Hum Genet* 68: 927–936

27 Risch N, Merikangas K (1996) The future of genetic studies of complex human diseases. *Science* 273: 1516–1517

28 Shiozawa S, Hayashi S, Tsukamoto Y, Goko H, Kawasaki H, Wada T (1998) Identification of the gene loci that predispose to rheumatoid arthritis. *Int Immunol* 10: 1891– 1895

29 Cornelis F, Faure S, Martinez M, Prud'homme JF, Fritz P (1998) New susceptibility locus for rheumatoid arthritis suggested by genome-wide linkage study. *Proc Natl Acad Sci USA* 95: 10746–10750

30 Osorio Y Fortea J, Bukulmez H, Petit-Teixeira E, Michou L, Pierlot C, Cailleau-Moindrault S, Lemaire I, Lasbleiz S, Alibert O et al (2004) Dense genome-wide link-

age analysis of rheumatoid arthritis, including covariates. *Arthritis Rheum* 50: 2757–2765

31   Jawaheer D, Seldin MF, Amos CI, Chen WV, Shigeta R, Etzel C, Damle A, Xiao X, Chen D, Lum RF et al (2003) Screening the genome for rheumatoid arthritis susceptibility genes: a replication study and combined analysis of 512 multicase families. *Arthritis Rheum* 48: 906–916

32   Eyre S, Barton A, Shephard N, Hinks A, Brintnell W, MacKay K, Silman A, Ollier W, Wordsworth P, John S, Worthington J et al (2004) Investigation of susceptibility loci identified in the UK rheumatoid arthritis whole-genome scan in a further series of 217 UK affected sibling pairs. *Arthritis Rheum* 50: 729–735

33   Lander E, Kruglyak L (1995) Genetic dissection of complex traits: guidelines for interpreting and reporting linkage results. *Nat Genet* 11: 241–247

34   Davis S, Weeks DE (1997) Comparison of nonparametric statistics for detection of linkage in nuclear families: single-marker evaluation. *Am J Hum Genet* 61: 1431–1444

35   Biswas S, Papachristou C, Irwin ME, Lin S (2003) Linkage analysis of the simulated data - evaluations and comparisons of methods. *BMC Genet* 4 (Suppl 1): S70

36   Jawaheer D, Gregersen PK (2002) The search for rheumatoid arthritis susceptibility genes: a call for global collaboration. *Arthritis Rheum* 46: 582–584

37   Fisher SA, Lanchbury JS, Lewis CM (2003) Meta-analysis of four rheumatoid arthritis genome-wide linkage studies: confirmation of a susceptibility locus on chromosome 16. *Arthritis Rheum* 48: 1200–1206

38   Wise LH, Lanchbury JS, Lewis CM (1999) Meta-analysis of genome searches. *Ann Hum Genet* 63: 263–272

39   Kennedy GC, Matsuzaki H, Dong S, Liu WM, Huang J, Liu G, Su X, Cao M, Chen W, Zhang J et al (2003) Large-scale genotyping of complex DNA. *Nat Biotechnol* 21: 1233–1237

40   Evans DM, Cardon LR (2004) Guidelines for genotyping in genomewide linkage studies: single-nucleotide-polymorphism maps *versus* microsatellite maps. *Am J Hum Genet* 75: 687–692

41   John S, Shephard N, Liu G, Zeggini E, Cao M, Chen W, Vasavda N, Mills T, Barton A, Hinks A et al (2004) Whole-genome scan, in a complex disease, using 11,245 single-nucleotide polymorphisms: comparison with microsatellites. *Am J Hum Genet* 75: 54–64

42   Suzuki A, Yamada R, Chang X, Tokuhiro S, Sawada T, Suzuki M, Nagasaki M, Nakayama-Hamada M, Kawaida R, Ono M et al (2003) Functional haplotypes of PADI4, encoding citrullinating enzyme peptidylarginine deiminase 4, are associated with rheumatoid arthritis. *Nat Genet* 34: 395–402

43   Huizinga TW, Pisetsky DS, Kimberly RP (2004) Associations, populations, and the truth: recommendations for genetic association studies in Arthritis & Rheumatism. *Arthritis Rheum* 50: 2066–2071

44   Johnson GC, Esposito L, Barratt BJ, Smith AN, Heward J, Di Genova G, Ueda H,

Cordell HJ, Eaves IA, Dudbridge F et al (2001) Haplotype tagging for the identification of common disease genes. *Nat Genet* 29: 233–237

45 The International HapMap Consortium (2003) The International HapMap Project. *Nature* 426: 789–796

46 Botstein D, Risch N (2003) Discovering genotypes underlying human phenotypes: past successes for mendelian disease, future approaches for complex disease. *Nat Genet* 33, Suppl: 228–237

47 Ueda H, Howson JM, Esposito L, Heward J, Snook H, Chamberlain G, Rainbow DB, Hunter KM, Smith AN, Di Genova G et al (2003) Association of the T-cell regulatory gene CTLA4 with susceptibility to autoimmune disease. *Nature* 423: 506–511

48 Colhoun HM, McKeigue PM, Davey SG (2003) Problems of reporting genetic associations with complex outcomes. *Lancet* 361: 865–872

49 Hirschhorn JN (2005) Genetic approaches to studying common diseases and complex traits. *Pediatr Res* 57: 74R–7R

50 Freimer N, Sabatti C (2004) The use of pedigree, sib-pair and association studies of common diseases for genetic mapping and epidemiology. *Nat Genet* 36: 1045–1051

51 Ioannidis JP, Trikalinos TA, Ntzani EE, Contopoulos-Ioannidis DG (2003) Genetic associations in large *versus* small studies: an empirical assessment. *Lancet* 361: 567–571

52 Ioannidis JP, Ntzani EE, Trikalinos TA, Contopoulos-Ioannidis DG (2001) Replication validity of genetic association studies. *Nat Genet* 29: 306–309

53 Lohmueller KE, Pearce CL, Pike M, Lander ES, Hirschhorn JN (2003) Meta-analysis of genetic association studies supports a contribution of common variants to susceptibility to common disease. *Nat Genet* 33: 177–182

54 Ke X, Durrant C, Morris AP, Hunt S, Bentley DR, Deloukas P, Cardon LR (2004) Efficiency and consistency of haplotype tagging of dense SNP maps in multiple samples. *Hum Mol Genet* 13: 2557–2565

55 Ardlie KG, Kruglyak L, Seielstad M (2002) Patterns of linkage disequilibrium in the human genome. *Nat Rev Genet* 3: 299–309

56 Pritchard JK, Rosenberg NA (1999) Use of unlinked genetic markers to detect population stratification in association studies. *Am J Hum Genet* 65: 220–228

57 Ogura Y, Bonen DK, Inohara N, Nicolae DL, Chen FF, Ramos R, Britton H, Moran T, Karaliuskas R, Duerr RH et al (2001) A frameshift mutation in NOD2 associated with susceptibility to Crohn's disease. *Nature* 411: 603–606

58 Hugot JP, Chamaillard M, Zouali H, Lesage S, Cezard JP, Belaiche J, Almer S, Tysk C, O'Morain CA, Gassull M et al (2001) Association of NOD2 leucine-rich repeat variants with susceptibility to Crohn's disease. *Nature* 411: 599–603

59 Rebbeck TR, Spitz M, Wu X (2004) Assessing the function of genetic variants in candidate gene association studies. *Nat Rev Genet* 5: 589–597

60 Little J, Bradley L, Bray MS, Clyne M, Dorman J, Ellsworth DL, Hanson J, Khoury

M, Lau J, O'Brien TR et al (2002) Reporting, appraising, and integrating data on genotype prevalence and gene-disease associations. *Am J Epidemiol* 156: 300–310

61 van Aken J, van Bilsen JH, Allaart CF, Huizinga TW, Breedveld FC (2003) The Leiden Early Arthritis Clinic. *Clin Exp Rheumatol* 21: S100–S105

62 Stolt P, Bengtsson C, Nordmark B, Lindblad S, Lundberg I, Klareskog L, Alfredsson L; EIRA study group (2003) Quantification of the influence of cigarette smoking on rheumatoid arthritis: results from a population based case-control study, using incident cases. *Ann Rheum Dis* 62: 835–841

63 Ahmad T, Marshall S, Jewell D (2003) Genotype-based phenotyping heralds a new taxonomy for inflammatory bowel disease. *Curr Opin Gastroenterol* 19: 327–335

64 Lee AT, Li W, Liew A, Bombardier C, Weisman M, Massarotti EM, Kent J, Wolfe F, Begovich AB, Gregersen PK (2005) The PTPN22 R620W polymorphism associates with RF positive rheumatoid arthritis in a dose-dependent manner but not with HLA-SE status. *Genes Immun* 6: 129–133

65 Criswell LA, Pfeiffer KA, Lum RF, Gonzales B, Novitzke J, Kern M, Moser KL, Begovich AB, Carlton VE, Li W et al (2005) Analysis of families in the multiple autoimmune disease genetics consortium (MADGC) collection: the PTPN22 620W allele associates with multiple autoimmune phenotypes. *Am J Hum Genet* 76: 561–571

66 Begovich AB, Carlton VE, Honigberg LA, Schrodi SJ, Chokkalingam AP, Alexander HC, Ardlie KG, Huang Q, Smith AM, Spoerke JM et al (2004) A missense single-nucleotide polymorphism in a gene encoding a protein tyrosine phosphatase (PTPN22) is associated with rheumatoid arthritis. *Am J Hum Genet* 75: 330–337

67 Bottini N, Musumeci L, Alonso A, Rahmouni S, Nika K, Rostamkhani M, MacMurray J, Meloni GF, Lucarelli P, Pellecchia M et al (2004) A functional variant of lymphoid tyrosine phosphatase is associated with type I diabetes. *Nat Genet* 36: 337–338

68 Viken MK, Amundsen SS, Kvien TK, Boberg KM, Gilboe IM, Lilleby V, Sollid LM, Forre OT, Thorsby E, Smerdel A, Lie BAV et al (2005) Association analysis of the 1858C>T polymorphism in the PTPN22 gene in juvenile idiopathic arthritis and other autoimmune diseases. *Genes Immun* 6: 271–273

69 Kyogoku C, Langefeld CD, Ortmann WA, Lee A, Selby S, Carlton VE, Chang M, Ramos P, Baechler EC, Batliwalla FM et al (2004) Genetic association of the R620W polymorphism of protein tyrosine phosphatase PTPN22 with human SLE. *Am J Hum Genet* 75: 504–507

70 Prokunina L, Castillejo-Lopez C, Oberg F, Gunnarsson I, Berg L, Magnusson V, Brookes AJ, Tentler D, Kristjansdottir H, Grondal G et al (2002) A regulatory polymorphism in PDCD1 is associated with susceptibility to systemic lupus erythematosus in humans. *Nat Genet* 32: 666–669

71 Kochi Y, Yamada R, Suzuki A, Harley JB, Shirasawa S, Sawada T, Bae SC, Tokuhiro S, Chang X, Sekine A et al (2005) A functional variant in FCRL3, encoding Fc

receptor-like 3, is associated with rheumatoid arthritis and several autoimmunities. *Nat Genet* 37: 478–485

72   Donn R, Alourfi Z, Zeggini E, Lamb R, Jury F, Lunt M, Meazza C, De Benedetti F, Thomson W, Ray D et al (2004) A functional promoter haplotype of macrophage migration inhibitory factor is linked and associated with juvenile idiopathic arthritis. *Arthritis Rheum* 50: 1604–1610

73   Donn RP, Shelley E, Ollier WE, Thomson W (2001) A novel 5'-flanking region polymorphism of macrophage migration inhibitory factor is associated with systemic-onset juvenile idiopathic arthritis. *Arthritis Rheum* 44: 1782–1785

Heterogeneity in rheumatoid arthritis based on expression analysis: towards personalised medicine

Cornelis L. Verweij and Tineke C.T.M. van der Pouw Kraan

Department of Molecular Cell Biology and Immunology, C270, VU Medical Centre, P.O. Box 7057, 1007 MB Amsterdam, The Netherlands

Introduction

The clinical presentation of patients with RA may reveal striking heterogeneity with a spectrum ranging from mild cases with a benign course to severe and erosive disease. In fact, heterogeneity is already reflected in the way patients are diagnosed with RA. Currently, the classifying diagnosis for RA is based on the presence of four out of seven criteria, which comprise a set of five clinical variables supplemented with radiographic evidence for erosions and the presence of rheumatoid factor (RF) as laboratory evidence [1, 2], which thus indicates that different sets of criteria are applied to classify 'the same' disease.

Heterogeneity between patients with RA is also reflected at the level of the distribution of mononuclear cells in the synovial lesions, which reveals a remarkable patient-specific organisation level [3–5]. In about 25% of the RA patients cellular infiltrates in the synovial tissue show a high degree of organisation resembling structures normally observed in lymph nodes, comprising distinct T- and B-cell areas and a network of follicular dendritic cells (FDC) within the B cell area, referred to as ectopic or tertiary lymphoid structures. In the remainder of the patients the tissues do not contain FDCs and show either a diffuse lymphocytic infiltrate or an aggregated T- and B-cell infiltrate [6].

The wide variation in responsiveness to virtually any treatment modality in RA is consistent with the heterogeneous nature of the disease [7, 8]. For example, despite the highly beneficial effects of TNF-blocking in suppressing disease, clear efficacy appears to be limited to a subset of patients [7, 8]. Since only about one third of the patients show a dramatic beneficial response (ACR 50 criteria) this suggests that additional cytokines and/or pathways contribute to disease. The relative contribution of the different disease pathways may vary between patients and, perhaps, between different stages of disease. Similar observations have been made for other therapies like treatment with CTLA4Ig, which blocks the interaction of CD80/86 on antigen presenting cells with CD28 on T-cells, and B-cell ablation therapy [9, 10].

The Hereditary Basis of Rheumatic Diseases, edited by Rikard Holmdahl
© 2006 Birkhäuser Verlag Basel/Switzerland

Hence, the cumulative data provides evidence that distinct pathogenetic mechanisms contribute to disease in RA. The heterogeneity most likely has its origin in the multifactorial nature of the disease, whereby specific combinations of environmental factor(s) and a varying polygenic background are likely to influence not only susceptibility but also the severity and disease outcome.

Unfortunately, important criteria to make selections of patients for optimal treatment and research purposes are currently lacking. Such criteria would be helpful to improve the treatment success for the individual RA patient. Given the destructive nature of the disease it would be highly desirable to predict in an early stage the most beneficial treatment for each individual patient. If we rely solely on clinical or radiological manifestations we will probably be responding too late to modify treatment in order to maximise protection. Moreover, patient stratification might also be helpful to assign homogeneous groups of patients for genetic studies and to improve the likelihood to observe efficacy in clinical trials. Thus, we need to improve methods to identify different forms and/or phases of disease.

Protein biomarkers in serum and synovial fluid of patients with RA

By definition, nearly every aspect of a disease phenotype should be represented in the pattern of genes and/or proteins that are expressed in the patient. Hence, the identification of differentially expressed genes and proteins may provide a comprehensive molecular description of disease heterogeneity that contributes to the identification of diagnostic and prognostic markers to allow stratification of patients. Studies of the proteome, i.e., the quantitative description of all proteins present in a tissue, cell or body fluid at a specific stage and at a given time, have the premise to uncover novel diagnostic and/or prognostic markers for RA.

Autoantibodies

Classical approaches have pursued the use of autoantibodies in the serum as useful biomarkers in RA. The presence of autoantibodies is a common feature in autoimmune diseases like RA and the pattern of autoantibodies present is used to distinguish between diseases. For RA the presence of specific autoantibodies against antigens containing one or more citrulline residues, the so-called anti-cyclic citrulline peptide (CCP) antibodies, and rheumatoid factor (RF) are instrumental in the classifying diagnosis [2, 11]. The fact that subsets of patients with RA carry either one of the RF or anti-CCP autoantibodies, both or neither in their serum is consistent with the heterogeneous nature of RA. In approximately two-thirds of the RA patients RF is present. Evidence exists that RF positivity predicts more severe disease [12]. Recent studies have demonstrated the use of anti-CCP as a valuable serologic marker to diagnose RA early in its development and distinguish it from other non-erosive types of arthritis [13–17]. A combination of biomarkers that involve RF and anti-CCP could reflect different aspects of the disease process, and therefore might be useful for evaluating prognosis in individual patients with early rheumatoid arthritis [18].

The advent of high-throughput technologies that allow parallel profiling of hundreds and thousands of proteins and genes has dramatically transformed the search for genes and proteins that might be important in diagnosis and disease management. Recent developments in the production of miniaturised autoantigen arrays allows parallel detection of autoantibody specificities [19]. Such protein arrays enable profiling of the specificity of autoantibody responses against panels of peptides and proteins representing known autoantigens as well as candidate autoantigens. This technology holds the promise to define autoantibody signatures that define subsets of patients with different clinical disease subtypes and treatment responses.

Protein biomarkers

A number of proteins in serum and synovial fluid have been studied on a one-by-one basis as biomarkers in RA. Among these, serum proteins such as C-reactive protein (CRP) offer good correlation with concurrent disease activity. Novel developments in protein chemistry and detection contributed considerable to global protein profiling in body fluids for protein marker discovery (Tab. 1).

Initial studies revealed that the heterodimer complex of the small calcium binding proteins S100A8 (calgranulin A/myeloid-related protein/MRP8) and S100A9 (calgranulin B/MRP14) was found to be increased in serum and SF of RA patients when compared to samples from patients with osteoarthritis and healthy controls [20]. S100A8 and S100A9 belong to a new class of inflammatory mediators, and function as a heterodimer [21]. These proteins are released by activated monocytes upon interaction with activated endothelial cells under inflammatory conditions. One of the functions of the heterodimer complex is to mediate leukocyte migration and adhesion to vascular endothelium. Liao et al. used tandem mass spectrometry (MS/MS), coupled with multidimensional liquid chromatography (LC) to identify biomarkers of disease severity in the synovial fluid and serum of patients with RA [22]. Significantly, levels of CRP, S100A8, S100A9 and S100A12 (calgranulin C) were elevated in the serum of patients with erosive disease compared with patients with non-erosive RA. Preliminary studies suggest that plasma levels of the S100A8/S100A9 heterocomplex are a useful marker in monitoring efficacy of anti-TNF-α therapy [20]. Hence the combined data suggest a correlation of a selective set of local and systemic markers of inflammation with disease type in RA.

Table 1 - Overview of genomics and proteomics studies to demonstrate heterogeneity in RA

Source: Serum/cell/ tissue type	No. of patients (RA)/healthy controls (HC) studied	Procedure: Mass spec./ microarray (# of genes)	Molecular heterogeneity Yes or No	Relation to clinical parameters	Biological process involved	Refs
PBMC	19 RA	4,300 genes	Yes	Early *vs.* established RA (11 *vs.* 8)	Immune/growth factor activity Proliferation/neoplasia	[26]
PBMC	14 RA	10,000 genes	No	RF$^+$ (6) *vs.* RF$^-$ (8)	–	[25]
PBMC	14 RA, 7 HC	10,000 genes	Yes (RA vs. HC)	–	Phagocytic function, inflammation (e.g., S100A8, S100A9)	[25]
Whole blood	25 RA	18,000 genes	Yes (between RA patients)	Unknown	Inflammation/ immune activity	Van der Pouw Kraan et al. (Unpublished observations)
	25 RA, 25 HC		Yes (RA vs. HC)	–	Inflammation/ immune activity	
Serum	10 RA	Mass spec.	Yes	Erosive *vs.* non-erosive (5 *vs.* 5)	Inflammation (e.g., CRP and S100 family members)	[20, 22] [27, 28]
Synovial tissue	23 RA	11,500/18,000 genes	Yes	ESR	Adaptive immunity (T-, and B-cells and APC), ectopic lymph nodes, STAT-1 pathway, tissue remodelling	
FLS	19 RA	18,000 genes	Yes	Unknown	TGF-β/Activin A pathway, myofibroblast differentiation, IGF2/IGFBP5	[36]

Abbreviations: FLS, fibroblasts like synoviocytes; PBMC, Peripheral blood mononuclear cells; ESR, erythrocyte sedimentation rate

Gene expression profiling in cells and tissues of patients with RA

A powerful way to gain insight into the molecular complexity of cells and tissues has arisen from DNA microarray technology, which provides the opportunity to determine differences in expression of a large portion of the genes in the genome in search for genes that are differentially expressed among patients with clinically defined RA. By large-scale gene expression profiling in blood cells and tissues of patients with RA one can generate a molecular portrait of disease stage or subtype. The molecular portrait typically represents the contributions and interactions of numerous distinct cells and diverse factors that are associated with disease. The differentially expressed gene sets may then be used as disease classifier and as surrogate marker for the involvement for a particular biological pathway in disease (Tab. 1). Such an approach has been proven useful in cancer research for the identification of classifiers for disease outcome and metastasis and underlying pathways [23, 24].

Heterogeneity in peripheral blood cells

Since RA is a systemic disease, several investigators study gene expression levels in peripheral blood cells to address the question, whether disease heterogeneity is detectable from gene expression levels in peripheral blood cells (Tab. 1, Fig. 1).

Comparison of the gene expression profiles of peripheral blood mononuclear cells (PBMC) between 14 RA patients and seven healthy controls using DNA-microarrays containing ~10K genes, revealed only nine genes that were differentially expressed [25]. These genes also reflected changes in the immune/inflammatory responses in RA patients, including the calcium-binding protein S100A8 and S100A10, which were expressed at higher levels in RA patients. The aspect of heterogeneity was addressed by principle component analysis, from which analysis it was clear that there is at least a greater variability in gene expression in RA patients than in controls. Subsequent analysis to identify differences between RF positive and RF negative RA patients yielded no differences in this study.

Gene expression profiling in whole blood cells from RA patients revealed significant variation between RA patients that allows stratification of patients on the basis of distinct molecular characteristics (Tineke van der Pouw-Kraan, Carla Wijbrandts, Paul-Peter Tak en Cornelis Verweij, unpublished observations). Analysis of the differentially expressed gene clusters using gene set analysis, such as PLAGE (www.cbcb.duke.edu/pathways/) and Panther (www.panther.appliedbiosystems.com), indicated that distinct inflammatory processes are active in different subsets of patients. So far the clinical relevance of the molecular differences remains to be established.

Olsen et al. identified different gene expression levels in PBMC from early arthritis (disease duration less than 2 years), and established RA patients (with an aver-

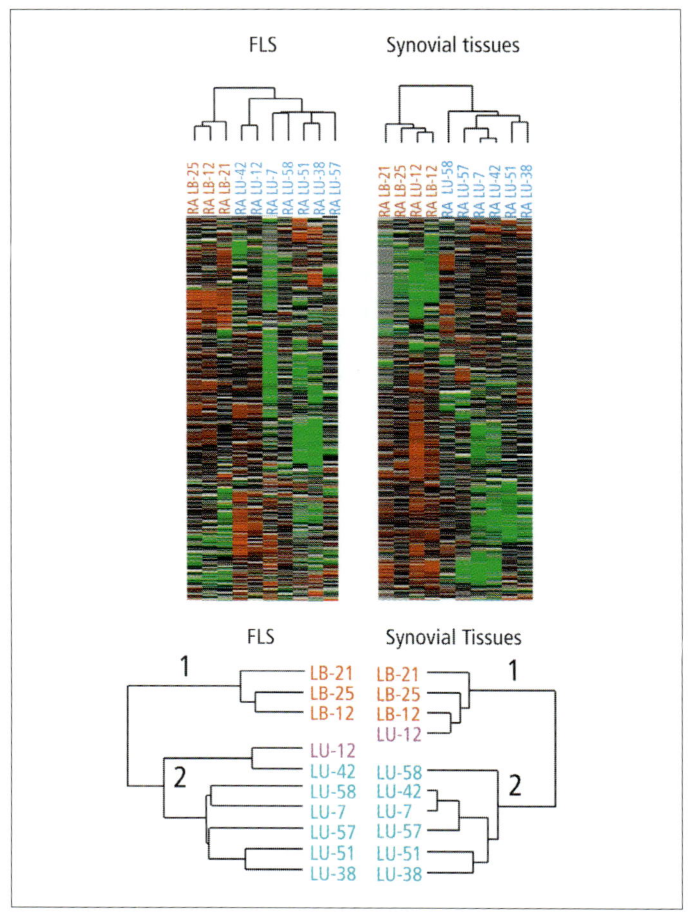

Figure 1

Heterogeneity between rheumatoid synovia is reflected in cultured fibroblast-like synovio-cytes (FLS) derived from the synovia.

A) Heterogeneity among FLS, and synovial tissue gene expression profiles. Hierarchical clus-tering of expression profiles from synovial tissues (left) and corresponding FLS (right) from 10 patients with rheumatoid arthritis (RA). Each row represents a single gene; each column represents an individual patient sample. Red bands indicate relative expression greater than the median of all samples (ratio >1), green bands indicate an expression level lower than the median (ratio <1), black bands indicate equal expression levels (ratio near 1), and grey bands indicate missing or excluded data. We selected genes whose transcripts varied in abundance by at least two-fold from their median level in at least two samples. B) Subclassification of synovial tissues matches the subclassification of FLS. Correlation between subclasses of FLS and corresponding synovial tissue samples.

age disease duration of 10 years) [26]. From 4,300 analysed genes, nine genes were expressed at three-fold higher levels, and 44 genes were expressed at three-fold lower levels in the early RA group compared to the late RA patients. The genes expressed at higher levels in early RA include colony stimulating factor 3 receptor, cleavage stimulation factor, and TGF-β receptor II, which affect B-cell function. Genes involved in immune/inflammatory processes and genes related to cell proliferation and neoplasia were expressed at lower levels in early arthritis. Comparison between the genes that were expressed at higher levels in early RA and influenza-induced genes indicated that about a quarter of the early arthritis genes overlapped with the influenza-induced genes. This finding led the authors to suggest that the early arthritis signature may partly reflect the response to an unknown infectious agent.

Heterogeneity in rheumatoid tissue revealed by gene expression profiling

Large-scale gene expression profiling of synovial tissues from patients with erosive RA (n = 23) using DNA microarrays with a complexity of ~11.5K genes (the so-called Lymphochip as described in [23, 27]) and ~18K genes [28] revealed considerable heterogeneity among different patients. A systematic characterisation of the differentially expressed genes highlighted the existence of at least two molecular distinct forms of RA tissues (Tab. 1, Fig. 1). One group revealed abundant expression of clusters of genes indicative of an ongoing inflammation and involvement of the adaptive immune response and is therefore referred to as high inflammation group. The expression of immunoglobulin genes showed the largest variation among RA patients with a high expression in the high inflammation RA tissues. Detailed analysis of the genes involved provide evidence for a prominent role of an activated STAT-1 pathway, including STAT-1 signalling receptors and STAT-1 target genes, among these STAT-1 itself [29]. Obviously, the various cytokines present in the RA synovium create a complex situation with simultaneous activation of multiple signalling pathways that may influence STAT-1 signalling. The importance of STAT activation in arthritis has been demonstrated in an animal model that periarticular administration of adenoviral suppressor of cytokine signalling 3 dramatically reduced the severity of collagen-induced arthritis and synovial IgG production [29]. These findings justify further research on the cell-specific expression of STAT signalling components, including the activating receptors and their ligands, which is crucial for our understanding of the molecular and cellular events that take place in the effector phase.

The expression profiles of the other group of RA tissues revealed a low inflammatory gene expression signature and an increased expression of genes involved in tissue remodelling activity, which is associated with fibroblast dedifferentiation. The gene expression signature of the latter tissues was reminiscent of that of tissues from patients with osteoarthritis. Careful analysis of the differentially expressed genes suggests the contribution of independent processes to disease and tissue destruction in RA.

The presence of the high inflammation tissues correlated with increased levels of systemic inflammation as is indicated by a significant increased erythrocyte sedimentation rate. However, the design of this study does not allow any firm conclusions to be drawn concerning the clinical parameters associated with the molecular phenotype. Therefore, further studies are necessary, which may provide a means to dissect and analyse the rheumatoid synovium of these patients and perform a thorough clinical association study based on molecular variation among patients.

Among the rheumatoid tissues that were grouped as high-inflammation tissues were tissues that contain organised lymph-node like follicular structures. Accordingly, the gene expression signature of these tissues revealed increased expression of 374 genes involved in, e.g., antigen presentation, chemotaxis and specific cytokine-induced activity. In contrast, tissues with a diffuse type of infiltrate showed a profile that indicated repression of angiogenesis and increased extracellular matrix remodelling. Based on the unique expression profile of tissues with a follicular infiltrate, we were able to predict the presence of lymphoid structures in rheumatoid synovia with a 100% correct classification in a ten-fold cross validation using PAM [30]. These results indicate that in tissues with ectopic lymphoid structures genes are overexpressed that reflect activation of distinct processes that allow attraction, retention and survival of FDC, T-, and B-cells with a high level of organisation. It is anticipated that patients with this type of tissue may require specific treatments to resolve the complex inflammatory processes that differs from that of the other types of synovial infiltrates.

Heterogeneity in rheumatoid fibroblast-like synoviocytes (FLS) revealed by gene expression profiling

Fibroblasts are ubiquitous mesenchymal cells that play important roles in organ development, inflammation, wound healing, fibrosis and pathology. Fibroblasts that are derived from different anatomic sites of the human body are differentiated in a site-specific fashion and are believed to play a key role in establishment and cellular organisation in tissues and organs, [31]. In chronic inflammation fibroblasts are considered as sentinel cells that contribute to leukocyte migration and local immune response through the production of various immune modulators [32]. These observations suggest these fibroblasts may acquire the capacity to modulate the immune response [33, 34].

Gene expression analysis of rheumatoid FLS identified two main groups. Most interestingly, hierarchical cluster analysis of paired synovial tissue and FLS samples revealed that the heterogeneity at the synovial tissue level is associated with a spe-

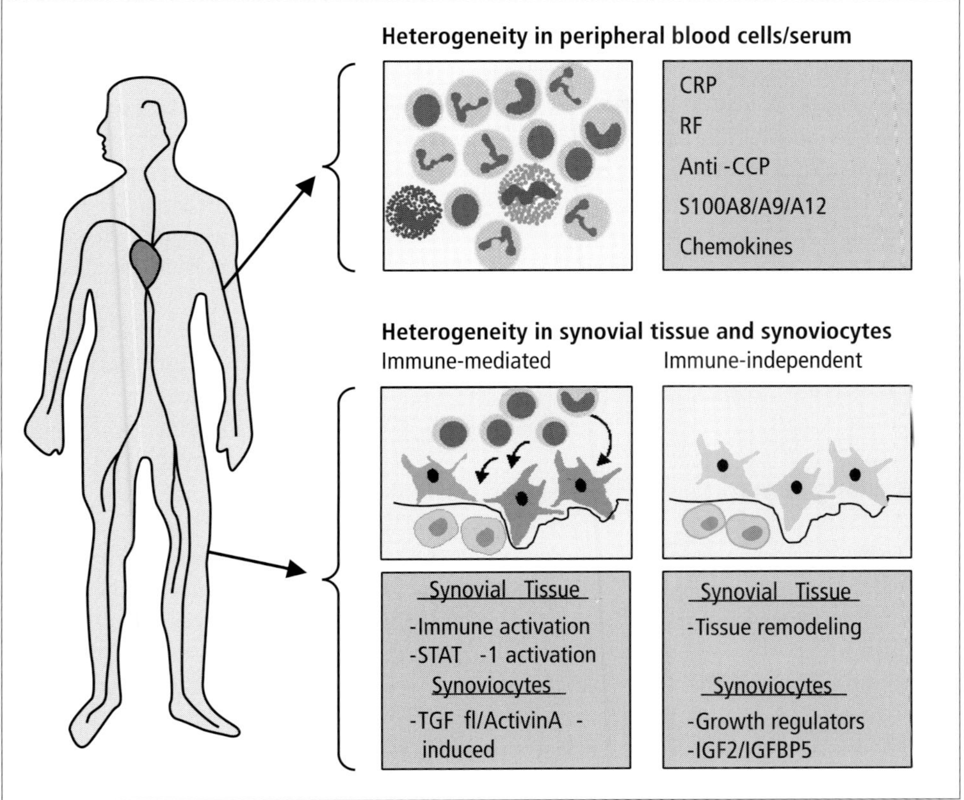

Figure 2
Schematic representation of heterogeneity in RA in different compartments of the human body

cific phenotypic characteristic of the resident fibroblast [32] (Figs 1 and 2). The high-inflammation tissues were associated with an FLS subtype that reveals similarity with so-called myofibroblasts. These cells are characterised by a marked increased expression of genes that represent the transforming growth factor β response program. Among these response genes were α-smooth muscle actin (SMA), SERPINE1, COL4A1 (type IV collagen-α chain), IER3 (immediate early response 3), and TAGLN (transgelin), and the gene for activin A as a potential agonist for the induction of this program. Increased growth factor (IGF2/IGFBP5) production appears to constitute a characteristic feature of FLS-derived of low inflammation tissues.

The molecular feature that defines the myofibroblast-like phenotype is reflected as an increased proportion of myofibroblast-like cells in the heterogeneous FLS population. Myofibroblast-like cells are also found upon immunohistochemical analysis of synovial tissues. The myofibroblast is a specialised fibroblast, which has acquired the capacity to express SMA, an actin isoform typical of vascular smooth muscle cells. It is now well accepted that the myofibroblast is a key cell for connective tissue remodelling and contributes to cell infiltration. Studies from the field of oncology indicate that myofibroblasts present in tumours play a crucial role in angiogenesis through the production of extracellular matrix proteins, chemokines and growth factors. Hence, it is proposed that myofibroblast-like synoviocytes in RA synovial tissue contributes to angiogenesis and ectopic lymph node development.

These data support the notion that heterogeneity between synovial tissues is reflected in the FLS as a stable trait and provide evidence for a link between an increased myofibroblast-like phenotype and high inflammatory synovitis.

Concluding remarks

The currently available data from genomics and proteomics studies in RA indicate that the observed molecular heterogeneity in RA mainly reflects differences in immune and inflammation activity (Tab. 1). Gene set analysis revealed differential activation of pathways such as the STAT-1 pathway in synovial tissue and the TGF-β/activin-α in the FLS. The existence of such molecular heterogeneity in RA that may be translated in distinct pathophysiological mechanisms at the site of the lesion would fit a model proposed by Firestein and Zvaifler [35], who suggested two independent processes in the destruction stage of RA. One is an immune mediated process that might progress to another phase that is centred on autonomous FLS aggression. This model is further supported by data from several animal models where FLS acquire a degree of independence from T-cell control in late destructive disease, implying an autonomous role for stromal elements, responsible for tissue destruction. Hence, both an immune-mediated and a stromal driven form of disease might drive destruction of bone and cartilage. How far this distinction relies on differences in the stage of the disease remains to be established. Clearly, large-scale gene- and protein expression profiling provides a molecular and biological basis for the well recognised but as yet poorly defined heterogeneity of RA.

The challenge now is to understand the biological basis underlying the differential gene expression programs between subgroups of RA patients. This will allow assignment of the biological pathways at play and may ultimately lead to the design of novel treatment modalities for a subgroup of patients. Moreover, we need to integrate the information on differential gene expression and protein patterns in RA with clinical parameters in such a way that it can be applied in a clinical setting to improve patient stratification and the accuracy of treatment decisions. Therefore, studies that provide solid evidence for the clinical relevance of the identified molecular differences have great priority for the direct future. The assignment of useful and reliable classifiers requires several levels of validation. It is anticipated that the classifying criteria comprise a constellation of biomarkers for use in a multianalyte approach that distinguish different forms or stages of RA. Ultimately, the identification of classifiers would be tremendously important to define homogeneous groups of patients for genetic and clinical studies, aimed to a more personalised therapy.

Acknowledgements

We are grateful to Drs. Pat Brown and David Botstein, in whose laboratories part of the work described in this report was performed.

Supported in part by the Howard Hughes Medical Institute, EU Marie Curie trainings network EURO-RA, and the Centre for Medical Systems Biology (a centre of excellence approved by the Netherlands Genomics Initiative, and grants from the National Cancer Institute, the Netherlands Organization for Scientific Research (NWO), and the Dutch Arthritis Foundation

References

1   Arnett FC, Edworthy SM, Bloch DA, McShane DJ, Fries JF, Cooper NS, Healey LA, Kaplan SR, Liang MH, Luthra HS et al (1988) The American Rheumatism Association 1987 revised criteria for the classification of rheumatoid arthritis. *Arthritis Rheum* 31: 315–324

2   Arnett FC (1989) Revised criteria for the classification of rheumatoid arthritis. *Bull Rheum Dis* 38: 1–6

3   Tak PP, Smeets TJ, Daha MR, Kluin PM, Meijers KA, Brand R, Meinders AE, Breedveld FC (1997) Analysis of the synovial cell infiltrate in early rheumatoid synovial tissue in relation to local disease activity. *Arthritis Rheum* 40: 217–225

4   Klimiuk PA, Goronzy JJ, Bjor nJ, Beckenbaugh RD, Weyand CM (1997) Tissue cytokine patterns distinguish variants of rheumatoid synovitis. *Am J Pathol* 151: 1311–1319

5   Ulfgren AK, Grondal L, Lindblad S, Khademi M, Johnell O, Klareskog L, Andersson U (2000) Interindividual and intra-articular variation of proinflammatory cytokines in patients with rheumatoid arthritis: potential implications for treatment. *Ann Rheum Dis* 59: 439–447

6   Takemura S, Braun A, Crowson C, Kurtin PJ, Cofield RH, O'Fallon WM, Goronzy JJ, Weyand CM (2001) Lymphoid neogenesis in rheumatoid synovitis. *J Immunol* 167: 1072–1080

7   Elliott MJ, Maini RN, Feldmann M, Kalden JR, Antoni C, Smolen JS, Leeb B, Breedveld FC, Macfarlane JD, Bijl H et al (1994) Randomised double-blind comparison of

chimeric monoclonal antibody to tumour necrosis factor alpha (cA2) *versus* placebo in rheumatoid arthritis. *Lancet* 344: 1105–1110

8   Elliott MJ, Maini RN, Feldmann M, Long-Fox A, Charles P, Katsikis P, Brennan FM, Walker J, Bijl H, Ghrayeb J et al (1993) Treatment of rheumatoid arthritis with chimeric monoclonal antibodies to tumor necrosis factor alpha. *Arthritis Rheum* 36: 1681–1690

9   Kremer JM, Westhovens R, Leon M, Di Giorgio E, Alten R, Steinfeld S, Russell A, Dougados M, Emery P, Nuamah IF et al (2003) Treatment of rheumatoid arthritis by selective inhibition of T-cell activation with fusion protein CTLA4Ig. *N Engl J Med* 349: 1907–1915

10 Edwards JC, Szczepanski L, Szechinski J, Filipowicz-Sosnowska A, Emery P, Close DR, Stevens RM, Shaw T (2004) Efficacy of B-cell-targeted therapy with rituximab in patients with rheumatoid arthritis. *N Engl J Med* 350: 2572–2581

11 Schellekens GA, de Jong BA, van den Hoogen FH, van de Putte LB, van Venrooij WJ (1998) Citrulline is an essential constituent of antigenic determinants recognized by rheumatoid arthritis-specific autoantibodies. *J Clin Invest* 101: 273–281

12 van Zeben D, Hazes JM, Zwinderman AH, Cats A, van der Voort EA, Breedveld FC (1992) Clinical significance of rheumatoid factors in early rheumatoid arthritis: results of a follow up study. *Ann Rheum Dis* 51: 1029–1035

13 Rantapaa-Dahlqvist S, de Jong BA, Berglin E, Hallmans G, Wadell G, Stenlund H, Sundin U, van Venrooij WJ (2003) Antibodies against cyclic citrullinated peptide and IgA rheumatoid factor predict the development of rheumatoid arthritis. *Arthritis Rheum* 48: 2741–2749

14 Nielen MM, van Schaardenburg D, Reesink HW, van de Stadt RJ, van der Horst-Bruinsma IE, de Koning MH, Habibuw MR, Vandenbroucke JP, Dijkmans BA (2004) Specific autoantibodies precede the symptoms of rheumatoid arthritis: a study of serial measurements in blood donors. *Arthritis Rheum* 50: 380–386

15 Berglin E, Padyukov L, Sundin U, Hallmans G, Stenlund H, van Venrooij WJ, Klareskog L, Dahlqvist SR (2004) A combination of autoantibodies to cyclic citrullinated peptide (CCP) and HLA-DRB1 locus antigens is strongly associated with future onset of rheumatoid arthritis. *Arthritis Res Ther* 6: R303–R308

16 Van Gaalen FA, Linn-Rasker SP, van Venrooij WJ, de Jong BA, Breedveld FC, Verweij CL, Toes RE, Huizinga TW (2004) Autoantibodies to cyclic citrullinated peptides predict progression to rheumatoid arthritis in patients with undifferentiated arthritis: a prospective cohort study. *Arthritis Rheum.* 50: 709–715

17 Bongi SM, Manetti R, Melchiorre D, Turchini S, Boccaccini P, Vanni L, Maggi E (2004) Anti-cyclic citrullinated peptide antibodies are highly associated with severe bone lesions in rheumatoid arthritis anti-CCP and bone damage in RA. *Autoimmunity* 37: 495–501

18 Lindqvist E, Eberhardt K, Bendtzen K, Heinegard D, Saxne T (2005) Prognostic laboratory markers of joint damage in rheumatoid arthritis. *Ann Rheum Dis* 64: 196–201

19 Robinson WH, DiGennaro C, Hueber W, Haab BB, Kamachi M, Dean EJ, Fournel S, Fong D, Genovese MC, de Vegvar HE et al (2002) Autoantigen microarrays for multiplex characterization of autoantibody responses. *Nat Med* 8: 295–301

20 Drynda S, Ringel B, Kekow M, Kuhne C, Drynda A, Glocker MO, Thiesen HJ, Kekow J (2004) Proteome analysis reveals disease-associated marker proteins to differentiate RA patients from other inflammatory joint diseases with the potential to monitor anti-TNFalpha therapy. *Pathol Res Pract* 200: 165–171

21 Roth J, Vogl T, Sorg C, Sunderkotter C (2003) Phagocyte-specific S100 proteins: a novel group of proinflammatory molecules. *Trends Immunol* 24: 155–158

22 Liao H, Wu J, Kuhn E, Chin W, Chang B, Jones MD, O'Neil S, Clauser KR, Karl J, Hasler F et al (2004) Use of mass spectrometry to identify protein biomarkers of disease severity in the synovial fluid and serum of patients with rheumatoid arthritis. *Arthritis Rheum* 50: 3792–3803

23 Alizadeh AA, Eisen MB, Davis RE, Ma C, Lossos IS, Rosenwald A, Boldrick JC, Sabet H, Tran T, Yu X et al (2000) Distinct types of diffuse large B-cell lymphoma identified by gene expression profiling. *Nature* 403: 503–511

24 van de Vijver MJ, He YD, van't Veer LJ, Dai H, Hart AA, Voskuil DW, Schreiber GJ, Peterse JL, Roberts C, Marton MJ et al (2002) A gene-expression signature as a predictor of survival in breast cancer. *N Engl J Med* 347: 1999–2009

25 Bovin LF, Rieneck K, Workman C, Nielsen H, Sorensen SF, Skjodt H, Florescu A, Brunak S, Bendtzen K (2004) Blood cell gene expression profiling in rheumatoid arthritis. Discriminative genes and effect of rheumatoid factor. *Immunol Lett* 93: 217–226

26 Olsen N, Sokka T, Seehorn CL, Kraft B, Maas K, Moore J, Aune TM (2004) A gene expression signature for recent onset rheumatoid arthritis in peripheral blood mononuclear cells. *Ann Rheum Dis* 63: 1387–1392

27 van der Pouw Kraan TC, Van Gaalen FA, Kasperkovitz PV, Verbeet NL, Smeets TJ, Kraan MC, Fero M, Tak PP, Huizinga TW, Pieterman E et al (2003) Rheumatoid arthritis is a heterogeneous disease: Evidence for differences in the activation of the STAT-1 pathway between rheumatoid tissues. *Arthritis Rheum* 48: 2132–2145

28 van der Pouw Kraan TC, Van Gaalen FA, Huizinga TW, Pieterman E, Breedveld FC, Verweij CL (2003) Discovery of distinctive gene expression profiles in rheumatoid synovium using cDNA microarray technology: evidence for the existence of multiple pathways of tissue destruction and repair. *Genes Immun* 4: 187–196

29 Lehtonen A, Matikainen S, Julkunen I (1997) Interferons up-regulate STAT1, STAT2, and IRF family transcription factor gene expression in human peripheral blood mononuclear cells and macrophages. *J Immunol* 159: 794–803

30 Tibshirani R, Hastie T, Narasimhan B, Chu G (2002) Diagnosis of multiple cancer types by shrunken centroids of gene expression. *Proc Natl Acad Sci USA* 99: 6567–6572

31 Chang HY, Chi JT, Dudoit S, Bondre C, van de RM, Botstein D, Brown PO (2002) Diversity, topographic differentiation, and positional memory in human fibroblasts. *Proc Natl Acad Sci USA* 99: 12877–12882

32 Smith RS, Smith TJ, Blieden TM, Phipps RP (1997) Fibroblasts as sentinel cells. Synthesis of chemokines and regulation of inflammation. *Am J Pathol* 151: 317–322

33 Brouty-Boye D, Pottin-Clemenceau C, Doucet C, Jasmin C, Azzarone B (2000) Chemokines and CD40 expression in human fibroblasts. *Eur J Immunol* 30: 914–919

34 Hogaboam CM, Steinhauser ML, Chensue SW, Kunkel SL (1998) Novel roles for chemokines and fibroblasts in interstitial fibrosis. *Kidney Int* 54: 2152–2159

35 Firestein GS, Zvaifler NJ (2002) How important are T cells in chronic rheumatoid synovitis?: II. T cell-independent mechanisms from beginning to end. *Arthritis Rheum* 46: 298–308

36 Kasperkovitz PV, Timmer TC, Smeets TJ, Verbeet NL, Tak PP, van Baarsen LG, Baltus B, Huizinga TW, Pieterman E, Fero M et al (2005) Fibroblast-like synoviocytes derived from patients with rheumatoid arthritis show the imprint of synovial tissue heterogeneity: Evidence of a link between an increased myofibroblast-like phenotype and high-inflammation synovitis. *Arthritis Rheum* 52: 430–441

Gene-based large scale LD-mapping of rheumatoid arthritis-associated genes

Ryo Yamada[1] and Kazuhiko Yamamoto[1,2]

[1]Laboratory for Rheumatic Diseases, SNP Research Center (E508), RIKEN, 1-7-22 Suehiro-cho, Tsurumi-ku, Yokohama, Kanagawa-ken 230-0045, Japan; [2]Department of allergy and rheumatology, Graduate school of medicine, University of Tokyo, Tokyo 113-8655, Japan

Introduction

For the last several years, linkage disequilibrium (LD) mapping using single nucleotide polymorphisms (SNP) has revealed disease-susceptible variants for many common disorders, including myocardial infarction [1, 2] diabetes mellitus [3] and osteoarthritis [4]. These findings have also been made for autoimmune diseases such as rheumatoid arthritis (RA) [5]. It is of particular interest that independent study groups have reported that several genes are associated with multiple autoimmune disorders [5]. Figure 1 summarises recent findings of disease-associated gene identification studies on autoimmune disorders and describes interrelations among multiple autoimmunities via disease-associated genes. In the following sections, the basics of SNP-based LD mapping and two major findings from the large-scale screening of RA-associated genes in a Japanese population are reviewed.

Gene-based LD mapping with SNPs

The genome project, SNPs, and the Hapmap project

Along with the completion of the human genome sequence in April 2003, numerous SNPs have been identified and registered and their number is still increasing (http://www.ncbi.nlm.nih.gov/projects/SNP/) [6]. In addition, a systematic project to establish a human polymorphism map, called the Hapmap project, is underway (http://www.hapmap.org/index.html) [7, 8]. It is important to make a large and detailed catalogue of SNPs and SNP-based haplotypes throughout the human genome, because these are the basic resources needed to facilitate studies on disease-associated loci for common and complex genetic disorders, including RA.

The Hereditary Basis of Rheumatic Diseases, edited by Rikard Holmdahl
© 2006 Birkhäuser Verlag Basel/Switzerland

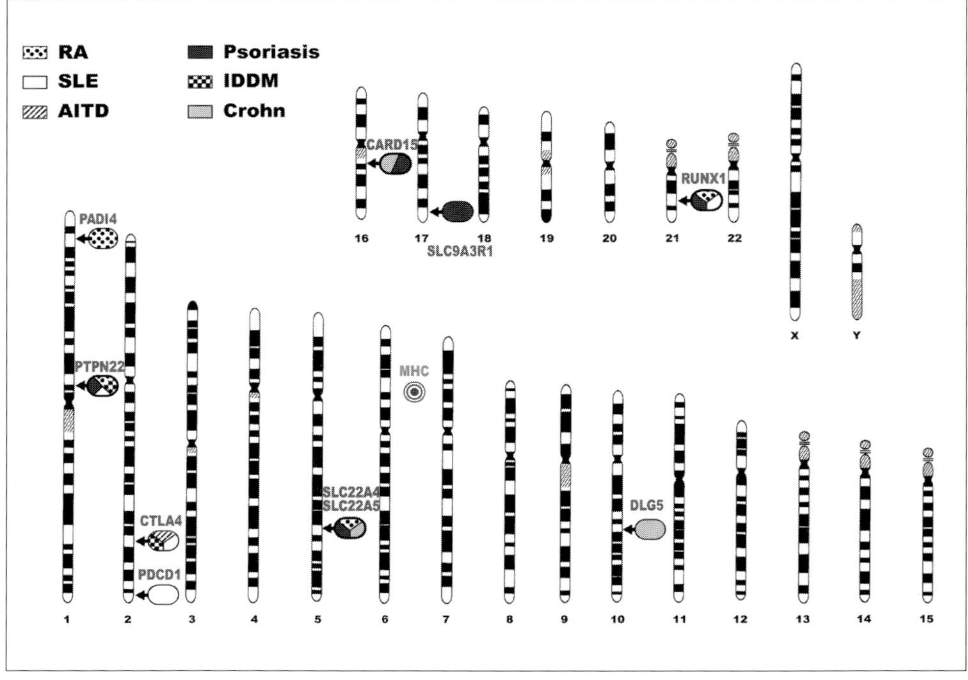

Figure 1

Multiple genes have been reported to have autoimmune disease-associated variant(s) and some of the genes were related to multiple autoimmune disorders. Some genes were associated with multiple autoimmune diseases. (a) The location of autoimmune disease-associated genes was plotted with an ideogram. (b) The inter-relation between genes and diseases and their crossovers.

RA, rheumatoid arthritis; SLE, systemic lupus erythematosus; AITD, autoimmune thyroid diseases; IDDM, insulin-dependent (type 1) diabetes mellitus.

Hypothesis-free and large-scale LD mapping

There is much left to be elucidated before the complete pathologic mechanism of RA can be understood. Thus all genes should be evaluated for their relation to RA without narrowing the range of target genes based on known but limited knowledge of the function or location of genes, that is, without a hypothesis in advance ('hypothesis-free'). Whole genome mapping studies including linkage analyses using familial cases and LD analyses with familial or non-familial subjects are designed to evaluate genes in a hypothesis-free fashion. In order to scan the whole genome in a hypothesis-free manner, large-scale screening must be performed and many genetic markers must be analysed. Because SNPs are the most abundant genetic markers throughout the human genome, and because methods to genotype them are simple and suitable for high-throughput systems, they are the best choice for common disease-associated gene mapping [9].

There are two ways to cover the genome with SNPs (Fig. 2). One way is to distribute SNPs evenly throughout the segment (the map-based approach) and the other is to distribute SNPs selectively or close to gene-coding regions (the gene-based approach). Although the number of SNPs needed to cover the whole genome depends on the population and disease to be analysed and the type of tests to be applied, roughly 500,000 to 1,000,000 SNPs would be necessary to cover the human whole genome with the map-based approach; the number of SNPs needed would be one-tenth that with the gene-based approach. The Hapmap project set its goal to register SNPs in the map-based fashion and RIKEN launched a gene-based mapping project with multiple reports for common diseases [1–4, 10, 11] (Figs 2a and 2b).

Identification of RA-associated variants in *PADI4* and *SLC22A4*

Features of screened chromosomes and genes: SNPs, LD, and haplotypes

We have performed large-scale, case-control LD mapping covering 22 autosomal chromosomes with the gene-based approach by using 100,926 SNPs that have been newly described in the Japanese population (http://snp.ims.u-tokyo.ac.jp/) [12]. Consequently, in the 1p domain, peptidylarginine deiminase type 4 (*PADI4*) was identified as an RA-related gene [13] and *SLC22A4* in 5q was reported to have RA-associated variants [14].

Gene-based LD mapping of 22 autosomal chromosomes (2,866 Mb; 93.4% of whole genome) domain was performed using 69,434 SNPs. It was revealed that those SNPs consisted of 7,875 LD blocks spanning 221.0 Mb, and 20,663 common haplotypes (frequency more than 5%) were inferred within blocks, that were tagged [15] with 12,441 htSNP. Among 21,153 genes annotated in autosomal chromosomes, 12,890 (60.9%) were screened with one or more SNPs and/or blocks. Although the genes that were evaluated with SNP(s) but were not covered by any block amounted to 4,509, half of them were screened with multiple SNPs. It is of note that number of investigated blocks was less than half that of genes, indicating the efficiency of block-based LD mapping (Fig. 3).

Identification of RA-associated SNPs

Initially we observed allele frequency and genotype distribution of all the registered SNPs, and LD extension was evaluated in Japanese controls. These data were refer-

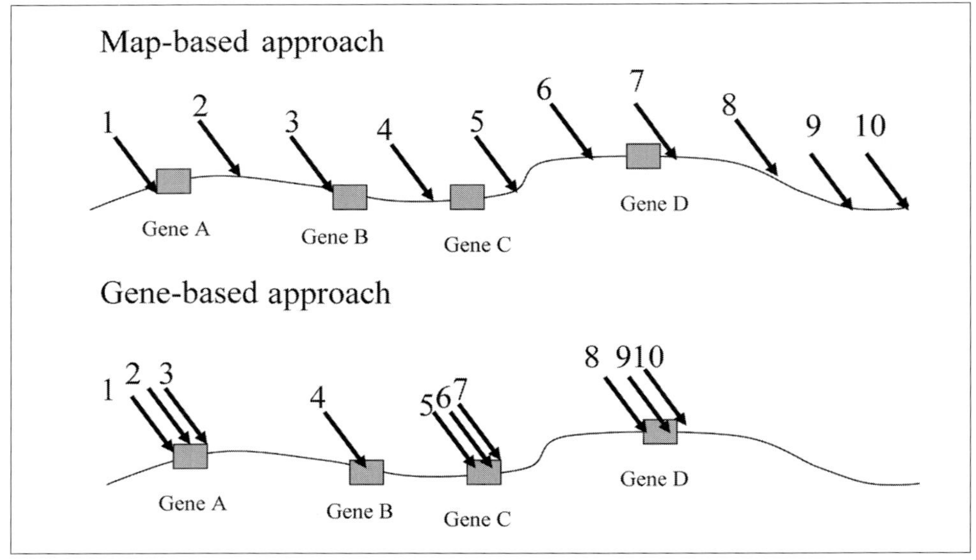

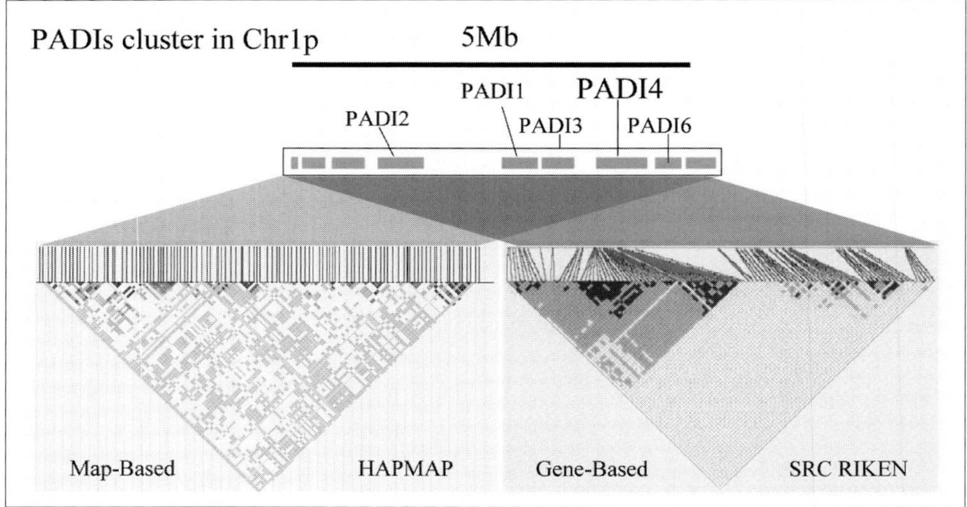

Figure 2

(a) Map-based selection and gene-based selection of SNPs. SNPs are evenly distributed on the genome without consideration of the presence of coding genes in the map-based selection. SNPs are densely distributed in gene-encoding regions in the gene-based selection.
(b) Distribution of SNPs and their LD pattern in the PADI gene-cluster was displayed in the map-based selection of SNPs from HapMap project data (left) and in gene-based selection from RIKEN's project (right).

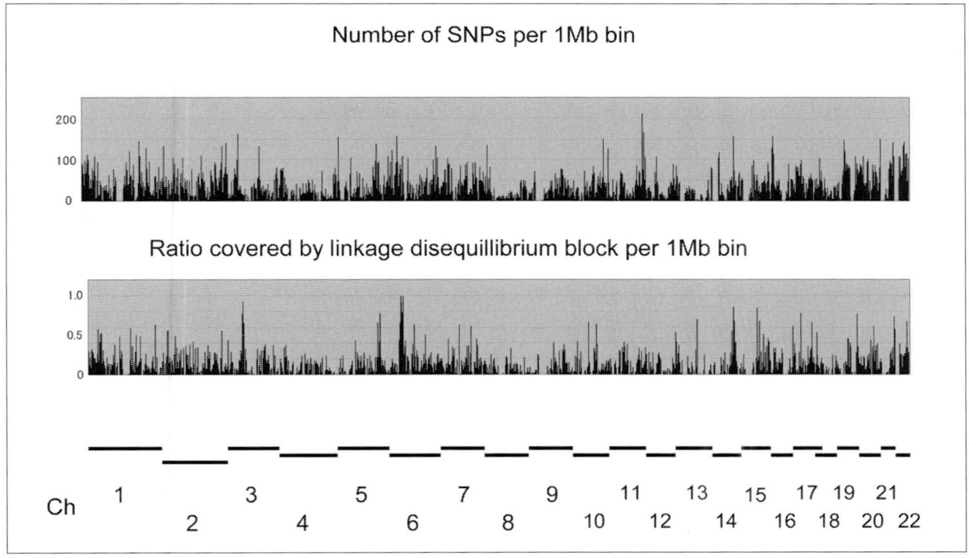

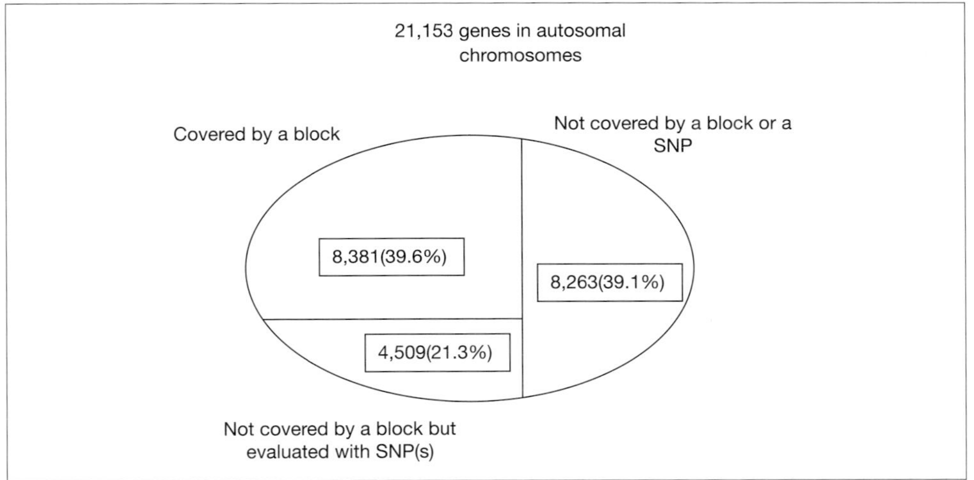

Figure 3
(a) Among 21,153 annotated genes by NCBI (build 34), we screened 12,890 (60.9%) of them. Majority of them were evaluated with multiple SNPs. (b) Because our project has been designed along with progress in human genome project and in listing of SNPs, and also because gene density varies throughout the genome, coverage of the genome in our project was uneven throughout the genome. The horizontal axis indicates tandem arrangement of 22 chromosomes. The upper histogram shows number of investigated SNPs per 1 Mb bin. The lower histogram indicates ratio covered by LD block for each 1 Mb bin.

enced as Japanese control information. To screen RA-associated genes, two types of case-control comparisons were designed. One was to compare diploid genotype data of individual SNPs between case and control groups. The other was to compare haplotype distributions of individual LD blocks between cases and controls. The former comparison strategy of individual SNPs identified RA-associated SNPs and further detailed analyses on their vicinity as well as identifying RA-associated polymorphisms in *PADI4* [11] and *SLC22A4* [10]. The latter strategy to compare haplotypes is now in progress. Both comparisons between case and control groups were initiated by genotyping limited numbers of RA cases and the preliminary data were compared to the Japanese control panel and SNPs or LD segments. Significant differences between the control panel and the RA cases were further validated with an increased number of cases (846 cases *versus* 658 controls). After identification of RA-associated SNP(s), we determined the extension of the LD in the loci and discovered polymorphisms in the LD segment by directly sequencing the loci with multiple subjects to obtain a denser SNP map. With the denser map, the origin of association was investigated.

PADI4

Identification of RA-associated polymorphisms

After identification of RA-associated SNPs in chromosome 1p36.1, where 5 PADI enzyme-encoding genes cluster in a 350 kb segment, detailed evaluation revealed that the association originated from a LD block containing a *PADI4* gene that has an RA-susceptible haplotype (Fig. 4a). The *PADI4* gene has two major haplotypes consisting of four SNPs. The relative risk of RA in individuals with two copies of the susceptible haplotype is 1.97 compared with individuals without a copy of the susceptible haplotype [13]. Because transcription from a susceptible haplotype is more stable than the other common haplotype of the *PADI4* gene, it is hypothesised that increased activity of *PADI4* produces susceptibility to RA. The association between the haplotypes and RA was not validated in a study in a British population, and further replication studies using multiple ethnic groups with more detailed evaluation of variants in the region would be beneficial [16].

Overview of RA, anti-citrullinated peptide antibodies, and PADI

The fact that autoantibody production against citrullinated proteins is highly specific to RA [17] and that *PADI4*, a gene encoding one of the citrullinating enzymes, is associated with RA [13] makes us believe that citrullination by PADI is a fundamental phenomenon in RA.

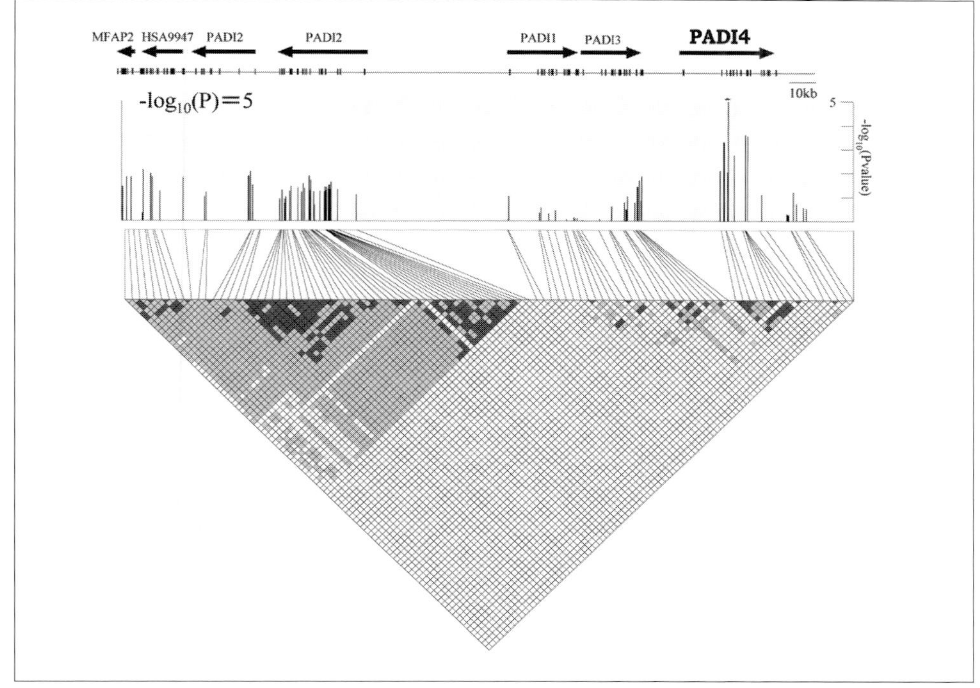

a

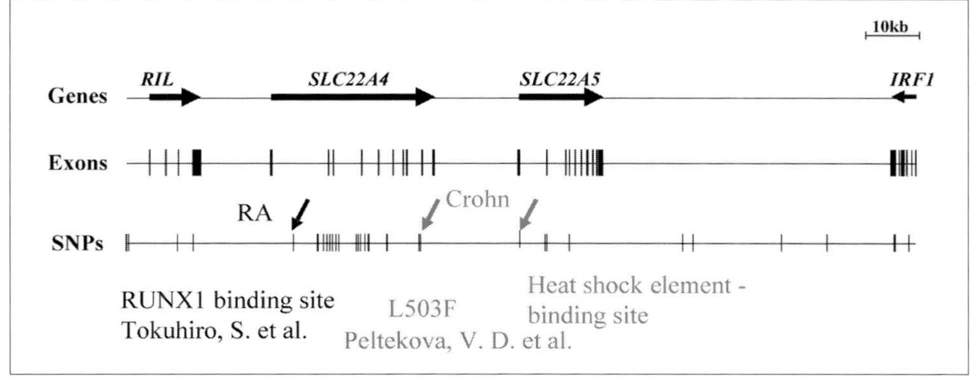

b

Figure 4
(a) Dense SNP map and result of case-control association test in the PADI gene-cluster. The peak of association was detected in PADI4 and the locus was separated into two LD blocks. (b) SLC22A4 and SLC22A5 are located next to each other. In Japanese, intronic SNPs in SLC22A4 were associated with RA and Crohn's disease. In Caucasian, haplotypes containing a missense SNP in SLC22A4 and a promoter SNP in SLC22A5 were associated with Crohn's disease.

Anti-citrullinated peptide antibodies in RA

Although various autoantibodies can be detected in the sera of patients with RA, several of these autoantibodies have been reported to be more specific and to have a higher positive predictive value for RA than others. Many highly RA-specific autoantibodies, such as antiperinuclear factor [18] anti-keratin antibody [19–21] and anti-Sa antibody [22] have been found to recognise citrullinated peptides [23–27]. Currently enzyme-linked immunosorbent assays to detect anti-citrullinated peptide antibody are considered one of the best diagnostic tests [28–32]. This assay is not only specific for RA (up to 98%), but the production of these autoantibodies can be detected very early in the disease – even several years before disease onset – and titre values tend to correlate with an erosive subtype of RA [25, 28–34].

Citrulline and peptidyl arginine deiminase enzymes

Citrulline and arginine. Citrulline is a non-coding native amino acid. It is a deiminated form of arginine. The biggest difference between arginine and citrulline is that arginine is one of the most basic amino acids and citrulline lacks the charged feature. The enzyme responsible for the conversion of peptidyl arginine to peptidyl citrulline is peptidyl arginine deiminase (PADI) [35]. Although some biological events, such as inflammation, apoptosis, trauma, ageing, and gene expression regulation by histones have been reported to be associated with post-translational citrullination [36, 37] the precise physiologic role of citrullination is still unknown [38–41]. Although the physiologic role of citrulline in peptides is not clear, what is known about citrullination and its consequences is that various proteins are citrullinated with subsequent changes in their conformation that lead to breakdowns in immunologic tolerance [33].

Enzymatic reaction and isozymes of PADIs. PADI enzymes catalyse the conversion of arginine residues to citrulline residues in proteins. Five isotypes of PADI – PADI1, 2, 3, 4, and 6 – have been cloned from several mammals, including humans [35]. All of these isotypes possess a Ca-binding motif, and they depend on high concentrations of Ca^{2+} for their enzymatic activity. The tissue distribution of the isotypes of PADIs varies. PADI2 and PADI4 were reported to be expressed in synovial tissue, and intranuclear localisation seems to characterise PADI4 [35].

PADIs in RA synovial tissue

PADI2 and PADI4

PADI2 and PADI4 were reported to be present in RA synovial tissue. As anticipated from the fact that PADI4 has an nuclear localisation signal but PADI2 does not, PADI4 was detected in the nucleus [35] and cytoplasm and PADI2 was detected only in cytoplasm. SNP-based whole genome surveys have identified RA-associated variants in PADI4 [13]. Increases in PADI4 activity due to polymorphisms in the gene seem to accelerate RA development [33]. Although one study from the United Kingdom did not report statistically significant results [16] it was not contradictory to the original study [13] and the association was replicated in another Japanese study [42]. A genetic contribution to RA of PADI4 variants, but not PADI2 variants, seems to be true, at least for the Japanese population, for the following reasons: all five PADI genes, including PADI2 and PADI4, cluster in chromosome 1p; genetic analysis discriminates PADI4 from PADI2; and the association with RA was only detected in PADI4 but not in PADI2.

Control of expression and activation of PADI2 and PADI4

PADI2 and PADI4 have different profiles of tissue distribution. PADI2 is relatively ubiquitously expressed and PADI4 is more specifically expressed in leukocytes [13]. The differences in tissue distribution, transcription, and translation of PADI2 and PADI4 are also regulated separately. Depending on cell lineages and their stages of differentiation, transcription and translation are regulated differently [43]. Moreover, enzymes of PADI2 and PADI4 are inactivated in steady state. Because a much higher concentration of intracellular Ca^{2+} is required for PADIs to be active as enzymes, regulation of Ca^{2+} seems to play a pivotal role in the regulation of PADI enzyme activation. A structural study of PADI4 revealed that binding of Ca^{2+} to PADI4 molecules produced catalytic sites in homodimers [44].

Newer findings on PADIs

Two mutually independent approaches, studies on autoantibodies and a gene-mapping study, highlighted citrulline in self-proteins as a breakthrough for understanding the precise pathophysiology of RA. For the last few years, newer insights on PADIs have been presented, such as the crystal structure of the PADI4 enzyme [44] and involvement of citrullination in a physiologic role in histone-related gene expression regulation [36, 45] along with expressional evidence on PADI isozymes, as well as validating data on the use of anti-CCP antibodies in clinics [31, 32]. Further investigations on citrulline and PADIs from various aspects will provide a more profound understanding of RA-related autoimmunity.

Solute carrier family 22 (organic cation transporter), member 4 (SLC22A4) and member 5 (SLC22A5)

Multiple autoimmune disorders are associated with SNPs in SLC22A4/A5 cluster in 5q31

The chromosome 5q31 cytokine cluster includes multiple T helper 2-type cytokines (the interleukin [IL] genes IL3, IL4, IL5, IL9, and IL13) as well as interferon regulatory factor-1 (IRF1), colony-stimulating factor-2 (CSF2), and T-cell transcription factor-7 (TCF7) (Fig. 4b). Initially, the association between RA and SNPs in SLC22A4 was reported [10], and a report on Crohn's disease-associated SNPs in SLC22A4 or SLC22A5 followed [46]. In addition, SLC22A4 was reported to have a functional variant associated with RA [14]. These reports indicated that SLC22A4/A5 locus has susceptible variants to multiple autoimmune inflammatory disorders and behaves in an ethnic-specific manner.

Identification of RA/Crohn's-associated polymorphisms in SLC22A4 and SLC22A5

As with *PADI4*, we initially identified RA-associated SNPs in a LD segment containing *SLC22A4* and *SLC22A5*. Further dense-LD mapping identified RA-associated SNPs (OR ~2.0) in the intronic region of SLC22A4 that disrupts the *RUNX1*-binding sequence [14]. Allele-specific effects of *RUNX1* binding on SLC22A4 expression seem to produce RA-susceptibility as in the case with *PDCD1* for systemic lupus erythematosus [47] and *SLC9A3R1* or *NAT9* for psoriasis [48, 49].

The whole genome linkage study identified IBD5 as a IBD-linked locus [50] and subsequent hierarchical strategy analysing trios using denser microsatellites narrowed the locus down to 1 cM with 2 loci. Then, the region was further closely evaluated with a dense SNP map that identified haplotypes consisting of two SNPs in SLC22A4 and SLC22A5 genes (OR ~1.6). The one SNP substitutes 503rd L to F with non-conservative effects on the tertiary structure of SLC22A4, and the other disrupts the heat shock element in 5' UTR of SLC22A5 (Fig. 3). Pharmacological assays revealed that the polymorphic amino acid substitution of SLC22A4 affected the transporting function of the molecule with several changes in V_{max} and K_m for some potential transport compounds. The allelic difference of 5' UTR SNP in SLC22A5 was observed in the binding of nuclear factors and *in vitro* transcription assay [46]. It was not concluded whether one or both of the genes or SNPs was responsible for disease susceptibility. Interestingly the SNPs associated with Crohn's disease in Caucasian subjects were not polymorphic in Japanese, and RA-associated SNPs in Japanese were also associated with Crohn's disease in Japanese [51].

SLC22A4 and SLC22A5

SLC22A4 is an organic cation transporter and is also called organic cation transporter 1 (OCT1). Organic cation transporters are among a large family of solute carrier transporters that number more than 200 in humans. Three organic cation transporters, SLC22A1 (OCT1), SLC22A2 (OCT2), and SLC22A3 (OCT3), were initially isolated and thereafter two more, SLC22A4 (OCTN1) and SLC22A5 (OCTN2), were isolated as a new family. The first two, SLC22A1 and SLC22A2, seem to transport organic cations in the renal basolateral membrane in a potentially dependent fashion. SLC22A5 is present in various tissues, including kidney, skeletal muscle, heart, and placenta [52]. In the kidney, SLC22A5 is expressed at the apical membrane of the proximal tubular epithelial cells [53]. A mouse strain with a point mutation of the mouse counterpart of SLC22A5 exhibited abnormal distribution of carnitine; this point mutation was identified as being involved in a familial carnitine metabolic disorder [54]. SLC22A5 is a physiologic transporter of carnitine and an *in vitro* analysis revealed that SLC22A5 transports tetraethylammonium (TEA) and carnitine. SLC22A4 and SLC22A5 are homologous at 76% (amino acids), and an *in vitro* investigation of SLC22A4 indicated that it transports carnitine as well as TEA. However the efficiency of carnitine transportation by SLC22A4 was far less than SLC22A5, and it is unlikely that carnitine is a physiologic substrate of SLC22A4. Very recently, a physiologic substrate of SLC22A4 was reported as ergothioneine.

Summary

Large-scale LD mapping has been successful at identifying RA-associated polymorphisms. Interestingly they seemed to identify RA-specific gene(s) and also gene(s) that contribute to multiple autoimmune diseases. Another important issue in the investigation of disease-associated polymorphisms is that polymorphisms vary among ethnic groups; therefore genetic studies should be carefully and extensively carried out with special attention paid to ethnic variations in polymorphisms and combinations of multiple genes.

Acknowledgements
We particularly thank Dr. Akari Suzuki and Dr. Shinya Tokuhiro for their dedicated work on PADI4 and SLC22A4, respectively, and Mr. Takahisa Kawaguchi and Mr. Hiroto Kawakami for their laborious assistance on genome wide statistics as well as all other members in the Laboratory for Rheumatic Diseases and other laboratories, SRC, RIKEN for our high-throughput LD-mapping.

References

1   Ozaki K, Ohnishi Y, Iida A, Sekine A, Yamada R, Tsunoda T, Sato H, Sato H, Hori M, Nakamura Y, Tanaka T (2002) Functional SNPs in the lymphotoxin-alpha gene that are associated with susceptibility to myocardial infarction. *Nat Genet* 32: 650–654
2   Ozaki K, Inoue K, Sato H, Iida A, Ohnishi Y, Sekine A, Sato H, Odashiro K, Nobuyoshi M, Hori M et al (2004) Functional variation in LGALS2 confers risk of myocardial infarction and regulates lymphotoxin-alpha secretion *in vitro*. *Nature* 429: 72–75

3   Kanazawa A, Tsukada S, Sekine A, Tsunoda T, Takahashi A, Kashiwagi A, Tanaka Y, Babazono T, Matsuda M, Kaku K et al (2004) Association of the gene encoding wingless-type mammary tumor virus integration-site family member 5B (WNT5B) with type 2 diabetes. *Am J Hum Genet* 75: 832–843

4   Kizawa H, Kou I, Iida A, Sudo A, Miyamoto Y, Fukuda A, Mabuchi A, Kotani A, Kawakami A, Yamamoto S et al (2005) An aspartic acid repeat polymorphism in asporin inhibits chondrogenesis and increases susceptibility to osteoarthritis. *Nat Genet* 37: 138–144. Epub 2005 Jan 9

5   Yamada R, Ymamoto K (2005) Recent findings on genes associated with inflammatory disease. *Mutat Res* 573: 136–151

6   Sherry ST, Ward MH, Kholodov M, Baker J, Phan L, Smigielski EM, Sirotkin K (2001) dbSNP: the NCBI database of genetic variation. *Nucl Acids Res* 29: 308–311

7   Gregersen PK (2003) Teasing apart the complex genetics of human autoimmunity: lessons from rheumatoid arthritis. *Clin Immunol* 107: 1–9

8   Laitinen T, Polvi A, Rydman P, Vendelin J, Pulkkinen V, Salmikangas P, Makela S, Rehn M, Pirskanen A, Rautanen A et al (2004) Characterization of a Common Susceptibility Locus for Asthma-Related Traits. *Science* 304: 300–304

9   Newton-Cheh C, Hirschhorn JN (2005) Genetic association studies of complex traits: design and analytical issues. *Mutat Res* 573: 54–69

10   Tokuhiro S, Yamada R, Chang X, Suzuki A, Kochi Y, Sawada T, Suzuki M, Nagasaki M, Ohtsuki M, Ono M et al (2003) An intronic SNP in a RUNX1 binding site of SLC22A4, encoding an organic cation transporter, is associated with rheumatoid arthritis. *Nat Genet* 35: 341–348

11   Suzuki A, Yamada R, Chang X, Tokuhiro S, Sawada T, Suzuki M, Nagasaki M, Nakayama-Hamada M, Kawaida R, Ono M et al (2003) Functional haplotypes of PADI4, encoding citrullinating enzyme peptidylarginine deiminase 4, are associated with rheumatoid arthritis. *Nat Genet* 34: 395–402

12   Haga H, Yamada R, Ohnishi Y, Nakamura Y, Tanaka T (2002) Gene-based SNP discovery as part of the Japanese Millennium Genome Project: identification of 190,562 genetic variations in the human genome. Single-nucleotide polymorphism. *J Hum Genet* 47: 605–610

13   Suzuki A, Yamada R, Chang X, Tokuhiro S, Sawada T, Suzuki M, Nagasaki M, Nakayama-Hamada M, Kawaida R, Ono M et al (2003) Functional haplotypes of PADI4, encoding citrullinating enzyme peptidylarginine deiminase 4, are associated with rheumatoid arthritis. *Nat Genet* 34: 395–402

14   Tokuhiro S, Yamada R, Chang X, Suzuki A, Kochi Y, Sawada T, Suzuki M, Nagasaki M, Ohtsuki M, Ono M, Furukawa H et al (2003) An intronic SNP in a RUNX1 binding site of SLC22A4, encoding an organic cation transporter, is associated with rheumatoid arthritis. *Nat Genet* 35: 341–348

15   Sebastiani P, Lazarus R, Weiss ST, Kunkel LM, Kohane IS, Ramoni MF (2003) Minimal haplotype tagging. *Proc Natl Acad Sci USA* 100: 9900–9905

16   Barton A, Bowes J, Eyre S, Spreckley K, Hinks A, John S, Worthington J (2004) A func-

tional haplotype of the PADI4 gene associated with rheumatoid arthritis in a Japanese population is not associated in a United Kingdom population. *Arthritis Rheum* 50: 1117–1121

17   Yamada R, Suzuki A, Chang X, Yamamoto K (2003) Peptidylarginine deiminase type 4: identification of a rheumatoid arthritis-susceptible gene. *Trends Mol Med* 9: 503–508

18   Sondag-Tschroots IR, Aaij C, Smit JW, Feltkamp TE (1979) The antiperinuclear factor. 1. The diagnostic significance of the antiperinuclear factor for rheumatoid arthritis. *Ann Rheum Dis* 38: 248–251

19   Young BJ, Mallya RK, Leslie RD, Clark CJ, Hamblin TJ (1979) Anti-keratin antibodies in rheumatoid arthritis. *Br Med J* 2: 97–99

20   Vincent C, de Keyser F, Masson-Bessiere C, Sebbag M, Veys EM, Serre G (1999) Anti-perinuclear factor compared with the so called "antikeratin" antibodies and antibodies to human epidermis filaggrin, in the diagnosis of arthritides. *Ann Rheum Dis* 58: 42–48

21   Vincent C, Serre G, Lapeyre F, Fournie B, Ayrolles C, Fournie A, Soleilhavoup JP (1989) High diagnostic value in rheumatoid arthritis of antibodies to the stratum corneum of rat oesophagus epithelium, so-called 'antikeratin antibodies'. *Ann Rheum Dis* 48: 712–722

22   Despres N, Boire G, Lopez-Longo FJ, Menard HA (1994) The Sa system: a novel antigen-antibody system specific for rheumatoid arthritis. *J Rheumatol* 21: 1027–1033

23   Simon M, Girbal E, Sebbag M, Gomes-Daudrix V, Vincent C, Salama G, Serre G (1993) The cytokeratin filament-aggregating protein filaggrin is the target of the so-called "antikeratin antibodies," autoantibodies specific for rheumatoid arthritis. *J Clin Invest* 92: 1387–1393

24   Girbal-Neuhauser E, Durieux JJ, Arnaud M, Dalbon P, Sebbag M, Vincent C, Simon M, Senshu T, Masson-Bessiere C, Jolivet-Reynaud C et al (1999) The epitopes targeted by the rheumatoid arthritis-associated antifilaggrin autoantibodies are posttranslationally generated on various sites of (pro)filaggrin by deimination of arginine residues. *J Immunol* 162: 585–594

25   Schellekens GA, de Jong BA, van den Hoogen FH, van de Putte LB, van Venrooij WJ (1998) Citrulline is an essential constituent of antigenic determinants recognized by rheumatoid arthritis-specific autoantibodies. *J Clin Invest* 101: 273–281

26   Masson-Bessiere C, Sebbag M, Girbal-Neuhauser E, Nogueira L, Vincent C, Senshu T, Serre G (2001) The major synovial targets of the rheumatoid arthritis-specific antifilaggrin autoantibodies are deiminated forms of the alpha- and beta-chains of fibrin. *J Immunol* 166: 4177–4184

27   Senshu T, Akiyama K, Kan S, Asaga H, Ishigami A, Manabe M (1995) Detection of deiminated proteins in rat skin: probing with a monospecific antibody after modification of citrulline residues. *J Invest Dermatol* 105: 163–169

28   Schellekens GA, Visser H, de Jong BA, van den Hoogen FH, Hazes JM, Breedveld FC, van Venrooij WJ (2000) The diagnostic properties of rheumatoid arthritis antibodies recognizing a cyclic citrullinated peptide. *Arthritis Rheum* 43: 155–163

29   Kroot EJ, de Jong BA, van Leeuwen MA, Swinkels H, van den Hoogen FH, van't Hof

M, van de Putte LB, van Rijswijk MH, van Venrooij WJ, van Riel PL (2000) The prognostic value of anti-cyclic citrullinated peptide antibody in patients with recent-onset rheumatoid arthritis. *Arthritis Rheum* 43: 1831–1835

30 Visser H, le Cessie S, Vos K, Breedveld FC, Hazes JM (2002) How to diagnose rheumatoid arthritis early: a prediction model for persistent (erosive) arthritis. *Arthritis Rheum* 46: 357–365

31 Meyer O, Labarre C, Dougados M, Goupille P, Cantagrel A, Dubois A, Nicaise-Roland P, Sibilia J, Combe B (2003) Anticitrullinated protein/peptide antibody assays in early rheumatoid arthritis for predicting five year radiographic damage. *Ann Rheum Dis* 62: 120–126

32 Rantapaa-Dahlqvist S, de Jong BA, Berglin E, Hallmans G, Wadell G, Stenlund H, Sundin U, van Venrooij WJ (2003) Antibodies against cyclic citrullinated peptide and IgA rheumatoid factor predict the development of rheumatoid arthritis. *Arthritis Rheum* 48: 2741–2749

33 Yamada R, Suzuki A, Chang X, Yamamoto K (2005) Citrullinated proteins in rheumatoid arthritis. *Frontiers in Bioscience* 10: 54–60

34 Quinn MA, Green MJ, Conaghan P, Emery P (2001) How do you diagnose rheumatoid arthritis early? *Best Pract Res Clin Rheumatol* 15: 49–66

35 Vossenaar ER, Zendman AJ, van Venrooij WJ, Pruijn GJ (2003) PAD, a growing family of citrullinating enzymes: genes, features and involvement in disease. *Bioessays* 25: 1106–1118

36 Cuthbert GL, Daujat S, Snowden AW, Erdjument-Bromage H, Hagiwara T, Yamada M, Schneider R, Gregory PD, Tempst P, Bannister AJ, Kouzarides T (2004) Histone deimination antagonizes arginine methylation. *Cell* 118: 545–553

37 Wang Y, Wysocka J, Sayegh J, Lee YH, Perlin JR, Leonelli L, Sonbuchner LS, McDonald CH, Cook RG, Dou Y et al (2004) Human PAD4 regulates histone arginine methylation levels via demethylimination. *Science* 306: 279–283

38 van Stipdonk MJ, Willems AA, Amor S, Persoon-Deen C, Travers PJ, Boog CJ, van Noort JM (1998) T cells discriminate between differentially phosphorylated forms of alphaB-crystallin, a major central nervous system myelin antigen. *Int Immunol* 10: 943–950

39 Rathmell JC, Thompson CB (1999) The central effectors of cell death in the immune system. *Annu Rev Immunol* 17: 781–828

40 Piacentini M, Colizzi V (1999) Tissue transglutaminase: apoptosis versus autoimmunity. *Immunol Today* 20: 130–134

41 Hershko A, Ciechanover A (1998) The ubiquitin system. *Annu Rev Biochem* 67: 425–479

42 Ikari K, Kuwahara M, Nakamura T, Momohara S, Hara M, Yamanaka H, Tomatsu T, Kamatani N (2005) Association between PADI4 and rheumatoid arthritis: A replication study. *Arthritis Rheum* 52: 3054–3057

43 Vossenaar ER, Radstake TR, van der Heijden A, van Mansum MA, Dieteren C, de Rooij DJ, Barrera P, Zendman AJ, van Venrooij WJ (2004) Expression and activity of citrulli-

nating peptidylarginine deiminase enzymes in monocytes and macrophages. *Ann Rheum Dis* 63: 373–381

44 Arita K, Hashimoto H, Shimizu T, Nakashima K, Yamada M, Sato M (2004) Structural basis for Ca^{2+}-induced activation of human PAD4. *Nat Struct Mol Biol* 11: 777–783. Epub 2004 Jul 11

45 Wang Y, Wysocka J, Sayegh J, Lee YH, Perlin JR, Leonelli L, Sonbuchner LS, McDonald CH, Cook RG, Dou Y et al (2004) Human PAD4 regulates histone arginine methylation levels via demethylimination. *Science* 306: 279–283

46 Peltekova VD, Wintle RF, Rubin LA, Amos CI, Huang Q, Gu X, Newman B, Van Oene M, Cescon D, Greenberg G et al (2004) Functional variants of OCTN cation transporter genes are associated with Crohn disease. *Nat Genet* 36: 471–475. Epub 2004 Apr 11

47 Prokunina L, Castillejo-Lopez C, Oberg F, Gunnarsson I, Berg L, Magnusson V, Brookes AJ, Tentler D, Kristjansdottir H, Grondal G et al (2002) A regulatory polymorphism in PDCD1 is associated with susceptibility to systemic lupus erythematosus in humans. *Nat Genet* 32: 666–669

48 Helms C, Cao L, Krueger JG, Wijsman EM, Chamian F, Gordon D, Heffernan M, Daw JA, Robarge J, Ott J et al (2003) A putative RUNX1 binding site variant between SLC9A3R1 and NAT9 is associated with susceptibility to psoriasis. *Nat Genet* 35: 349–356

49 Yamada R, Tokuhiro S, Chang X, Yamamoto K (2004) SLC22A4 and RUNX1: identification of RA susceptible genes. *J Mol Med* 82: 558–564

50 Rioux JD, Silverberg MS, Daly MJ, Steinhart AH, McLeod RS, Griffiths AM, Green T, Brettin TS, Stone V, Bull SB et al (2000) Genomewide search in Canadian families with inflammatory bowel disease reveals two novel susceptibility loci. *Am J Hum Genet* 66: 1863–1870

51 Yamazaki K, Takazoe M, Tanaka T, Ichimori T, Saito S, Iida A, Onouchi Y, Hata A, Nakamura Y (2004) Association analysis of SLC22A4, SLC22A5 and DLG5 in Japanese patients with Crohn disease. *J Hum Genet* 49: 664–668

52 Tamai I, Ohashi R, Nezu JI, Sai Y, Kobayashi D, Oku A, Shimane M, Tsuji A (2000) Molecular and functional characterization of organic cation/carnitine transporter family in mice. *J Biol Chem* 275: 40064–40072

53 Tamai I, China K, Sai Y, Kobayashi D, Nezu J, Kawahara E, Tsuji A (2001) Na(+)-coupled transport of L-carnitine via high-affinity carnitine transporter OCTN2 and its subcellular localization in kidney. *Biochim Biophys Acta* 1512: 273–284

54 Nezu J, Tamai I, Oku A, Ohashi R, Yabuuchi H, Hashimoto N, Nikaido H, Sai Y, Koizumi A, Shoji Y et al (1999) Primary systemic carnitine deficiency is caused by mutations in a gene encoding sodium ion-dependent carnitine transporter. *Nat Genet* 21: 91–94

C. Shared heredity of rheumatic diseases

Emerging relationships: rheumatoid arthritis and the PTPN22 associated autoimmune disorders

Peter K. Gregersen[1] and Robert M. Plenge[2]

[1]Robert S. Boas Center for Genomics and Human Genetics, The Feinstein Institute for Medical Research, North Shore LIJ Health System, 350 Community Drive, Manhasset, New York 11030, USA; [2]Broad Institute of MIT and Harvard, Division of Rheumatology, Allergy and Immunology, Brigham and Women's Hospital, Cambridge, MA, USA

Introduction

The complexity and variety of autoimmune diseases and phenotypes is a source of continuing fascination and frustration for both physicians and scientists. Even in the setting of exciting advances in understanding the molecular basis of immune recognition and regulation, the fundamental causes of the autoimmune diseases are unknown, and it is still unclear to what extent these diseases are pathogenically related. The mere presence of autoimmune phenomena is not a compelling unifying feature, since these phenomena can occur in a variety of circumstances, including in normal subjects. Studies in animal models and the discovery of a few rare highly penetrant genes involved in human polyendocrine syndromes [1], have provided some direct evidence that abnormalities in thymic selection [2] can cause autoimmunity in humans. However, it is still not clear to what extent altered thymic selection operates in the more common autoimmune diseases [3], and other peripheral regulatory mechanisms are also likely to be important [4, 5]. Comprehensive genetic analysis in humans is one of the most promising avenues of gaining insight into this question. In the last few years, this approach has begun to bear fruit.

The intracellular tyrosine phosphatase, PTPN22, has recently been associated with a variety of autoimmune diseases, including type 1 diabetes [6], rheumatoid arthritis [7], systemic lupus [8] and autoimmune thyroid disease [9–11]. These discoveries have provided some of the first direct evidence that these diseases are genetically related, and also raise a host of new questions to be addressed. As discussed in this review, the example of PTPN22 illustrates how human genetics is likely to provide new insights into pathogenesis and unravel new relationships among the autoimmune diseases.

The Hereditary Basis of Rheumatic Diseases, edited by Rikard Holmdahl

Familial aggregation of autoimmune diseases and phenotypes

A common means of establishing the extent of the genetic contribution to a disease or phenotype is to examine the degree of familial aggregation. For Mendelian disorders, the pattern of familial aggregation often suggests a genetic model (dominant or recessive). However, for the common autoimmune diseases, the extent of familial aggregation is quite modest, large extended families are rare in the population, and the genetic model is unknown. A commonly accepted method of quantifying familial aggregation is to compare the prevalence of disease in relatives of affected individuals (e.g., siblings), with the prevalence in the general population. In the case of siblings, this value is designated the relative risk to siblings (λ_s) and is generally in the range of 10–20 for many autoimmune diseases [12]. In the case of rheumatoid arthritis, the λ_s is probably somewhat lower, in the range of 5–10 [13]. Although shared environmental factors among siblings can also explain a λ_s that is greater than 1, it is likely that a substantial proportion of the increased risk to siblings is due to the fact that siblings share genetic variation in common with the affected individuals in these families. In addition, for phenotypes that are dependent on complex combinations of relatively uncommon genotypes, there may be very low degrees of familial aggregation, even in the presence of a large genetic component to the phenotype [14].

In addition to this evidence for familial aggregation of particular disease phenotypes, it is also apparent that in some cases *different* autoimmune disorders also tend to aggregate together in families with one another. This issue is difficult to study since it requires the study of sizable populations of affected subjects with careful characterization of disease phenotypes in their relatives. The best evidence for familial aggregation across diseases involves type 1 diabetes, rheumatoid arthritis and autoimmune thyroid disease [15–17]. SLE probably also falls into this group [18, 19], as well as other less common autoimmune disorders [20]. Multiple sclerosis may also have some tendency toward familial aggregation with these disorders although the data on this is less clear [21–23].

With the exception of MS, an excess of autoantibodies in unaffected relatives is also well described for many of these diseases. For example, approximately 20% of the first degree relatives of patients with SLE have evidence of humoral autoimmunity to one or more nuclear antigens [24–26]. There is also evidence of increased thyroid autoimmunity and anti-thyroid antibodies in unaffected family members of SLE patients [27], as well as an increase in anti-thyroglobulin autoantibodies in the SLE patients themselves [28]. Antibodies to citrullinated peptides (anti-CCP) are now widely accepted as being quite specific for rheumatoid arthritis: anti-CCP Abs are present in up to 80% of RA patients but less than 1% in the general population [29, 30]. There is no published data concerning the rate of these antibodies in relatives of RA patients. However autoimmune thyroid disease is present in 15–20% of patients with RA [31] and the rate of thyroid autoantibodies in their unaffected first degree relatives is approximately double the rate expected in a control population [32].

Thus, the available data indicates that there is a significant tendency for some autoimmune phenotypes to aggregate with another in families, and this suggests that common genes may underlie these different disorders.

Genetic studies based on linkage suggest genetic overlap among different autoimmune diseases

Beginning in the 1990s many groups began to perform linkage studies in order to identify chromosomal regions that harbor risk genes for autoimmune diseases. In general, these studies took the form of affected sibling pair analyses, in which there is an attempt to identify genetic regions that are shared among affected siblings more frequently than expected by chance. This approach to genetic analysis is discussed in more detail by Worthington (see Chapter 2). In general, affected sibling pair analysis suffers from lack of statistical power, and therefore requires very large numbers of families in order to provide 'definite' evidence for linkage to a particular genetic region [33]. Nevertheless, the combined data have suggested that common genetic regions may be involved across different autoimmune disorders [34]. More recent data continue to suggest that lupus, type 1 diabetes and rheumatoid arthritis appear to have some areas of overlap involving particularly certain regions of chromosomes 1q, 2q, and 6q, among others [35–39]. However, the evidence of linkage in all cases is modest, and not always replicated. In addition, each of these linkage regions contain large numbers of potential candidate genes, and the genes involved in the different disorders may well be quite different, even within the same linkage region. Thus, while there was considerable early interest in these overlapping linkage results, it is not clear how significant these findings really are, and they do not in and of themselves provide a strong case for a common genetic basis for autoimmune disorders.

Genetic studies based on association: from candidate genes to 'genome-wide' association studies

Aside from the fact of familial aggregation, the earliest evidence of a genetic component to autoimmunity was the discovery of the HLA associations with these disorders. This is the classic example of 'candidate gene' association studies, in which a genetic variant is studied because it is located in a gene of obvious functional interest. The candidate gene approach has been widely applied in recent years for a variety of reasons. First, it is hypothesis driven, and thus conforms to traditional investigator initiated approaches to scientific investigation. Second, it generally requires a relatively small number of genotypes to be performed and thus is not overwhelmingly costly. Thirdly, advances in understanding the molecular basis of both the adaptive and innate immune response have presented a wealth of provocative candidate genes to study. Of course, linkage results may also drive the selection of candidate genes. In this case, the positional mapping information provides the underlying 'hypothesis' that a risk gene exists in a particular genetic region. Indeed, the end game of gene identification nearly always takes the form of a candidate gene association study, regardless of what method is used for the mapping.

Because of the rapid advances in genotyping technology, it is now possible to pursue a broad 'discovery driven' approach to genetic association studies. This may involve a focus on functional polymorphisms in a large number, or all, of the known genes. This approach is currently being taken by Celera Diagnostics, and applied in combination with knowledge of linkage data, led to the discovery of the PTPN22 association with rheumatoid arthritis [7, 40] (see below). Chip technology is now available to pursue a variant of this approach [41]. Alternatively, very comprehensive genome-wide association studies are fast becoming a reality, with the simultaneous testing of hundreds of thousands of SNPs across the genome using various platforms [42, 43]. The computational and statistical issues raised by these studies are substantial [44, 45], and in general will result in the production of hundreds or thousands of positive findings that will require confirmation in follow up studies. Nevertheless, early results suggest that genome-wide association is likely to be a very powerful means of identifying common susceptibility genes for autoimmunity and other complex disorders [46].

PTPN22: two routes to discovery

As discussed below, the importance of kinases and phosphatases for regulating T-cell signaling, as well as a wide variety of biological functions, has been an area of rapidly expanding research [47]. In early 2004, Bottini et al. reported the association of the intracellular tyrosine phosphatase PTPN22 with type 1 diabetes [6]. This discovery resulted from a candidate gene approach, informed by a detailed knowledge of the importance of phosphatases for regulating T-cell function. The associated polymorphism (rs2476601, 1858C→T) was found to confer a relative risk (RR) of ~1.8 for carriers of the susceptible allele, and the disease-associated allele results in an amino acid substitution of tryptophane (W) for arginine (R) at position 620. This amino acid change was located in one of four proline rich SH3 binding sites (Fig. 1), and was shown to disrupt the binding of PTPN22 to the intracellular kinase, Csk.

The PTPN22 association with rheumatoid arthritis was found using an alternative experimental approach based on a broad 'functional' genome-wide association study [7], informed in part by knowledge of the linkage results from affected sibling pair analysis [36, 40]. Again, positive associations with the 620W allele were

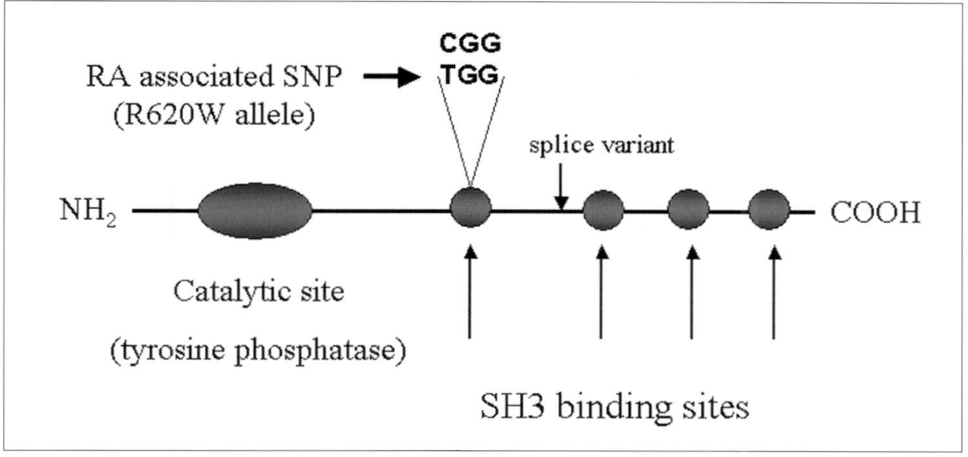

Figure 1
Domain organization of the PTPN22 protein showing the catalytic site in the N-erminal region, followed by four proline rich SH3 binding sites. The R620W polymorphism is located in the first SH3 binding site. Splice variants as indicated in the figure have been described [54].

observed in both a discovery and replication cohort of cases and controls, with RR ~1.7–2.0 overall. The effect of the risk allele on reducing binding of PTPN22 to Csk was also confirmed.

Thus, the finding of an association of PTPN22 with autoimmune disease resulted from two complementary approaches to gene discovery, one mainly hypothesis driven and the other largely discovery driven. It is likely that both of these experimental approaches will be successfully employed in the future to identify additional genes involved in autoimmunity.

PTPN22 is a negative regulator of early T-cell signaling events

Protein tyrosine phosphatase non-receptor type 22 (PTPN22) belongs to a group of intracellular tyrosine phosphatases, part of a larger family of over 100 phosphatases with diverse functions [48]. Approximately 30 phosphatases are expressed in T-cells, and tyrosine phosphatase activity is generally associated with a negative regulatory effect on T-cell function. Thus, early experiments showed that generalized phosphatase inhibition results in persistent proliferation of polyclonally activated T-cells, or can induce spontaneous activation and cytokine release by resting T-cells [49, 50]. A specific role of PTPN22 in T-cell regulation has been confirmed by the results of knocking out the murine homolog of PTPN22 (PEP), resulting in lowered thresholds for T-cell receptor signaling in these animals [51]. PEP knockout mice on a non-autoimmune background (C57/Bl6) exhibit a variety of phenotypes consistent with T-cell hyper-responsiveness, including enlargement of spleen and lymph nodes due to T-cell proliferation. This becomes more prominent in older mice, with the spontaneous development of germinal centers that appear to be largely dependent on the enhanced T-cell function present in the PEP$^{-/-}$ animals. Increased T-cell proliferative capacity is primarily found within the effector/memory cell compartment in both CD4 and CD8 subsets, and this is accompanied by enhanced phosphorylation of activating tyrosine residues in both Lck and ZAP70. Although there were increases in the levels of certain Ig isotypes in these knockout animals, autoantibodies did not develop, nor were there signs of overt autoimmune disease. Thus PEP deficiency alone does not lead to clinical autoimmunity. A negative effect on T-cell activation has also been demonstrated in human T-cells, where knockdown of PTPN22 expression using RNAi results in reduced NF-κB translocation after TCR signaling [7].

One site of PTPN22 action involves the regulation of Lck activation, as illustrated in Figure 2. As mentioned above, PTPN22 has been shown to bind to an intracellular tyrosine kinase, Csk [52, 53]. This binding occurs by virtue of a proline rich SH3 binding site on PTPN22, interacting with the SH3 domain of Csk. As shown on the right hand side of Figure 2, these molecules act in concert to inactivate Lck, a src family kinase that is involved in early T-cell signaling events. Csk acts to phosphorylate tyrosine 505 (an inhibitory phosphate for Lck), while PTPN22 acts to remove the activating phosphate at tyrosine 394. The combined effect of these activities is to convert Lck to an inactive configuration, as shown on the left side of the figure.

The PTPN22 R620W polymorphism is located within the SH3 binding site of PTPN22, and the tryptophan (W) substitution at this position has been shown to disrupt the binding of PTPN22 to Csk [6, 7]. Thus, one possible effect of the disease-associated 620W allele is a reduction in the dephosporylation of tyrosine 394 on Lck, and with reduced or delayed downregulation of T-cell receptor signaling. Clearly, this polymorphism does not completely eliminate the functions of PTPN22, since even homozygous carriers of PTPN22 620W do not exhibit a phenotype like the knockout mouse. It is more likely that the 620W polymorphism results in a change in the level of effective PTPN22 activity in particular cell compartments. Although the PEP knockout animals provide an intriguing phenotype of enhanced and/or persistent T-cell activation, it must be admitted that there is currently no direct evidence for a functional effect of the R620W on T-cell activation in humans. It is not even clear whether we should expect a loss of function or a gain of function. Conceivably, the reduced binding of the 620W allele to Csk may make PTPN22 more available for dephosphorylation of Lck or other target molecules, such as ZAP70. Indeed, there is evidence that PTPN22 also binds to c-Cbl, through which it is brought into contact with ZAP70 [47, 54].

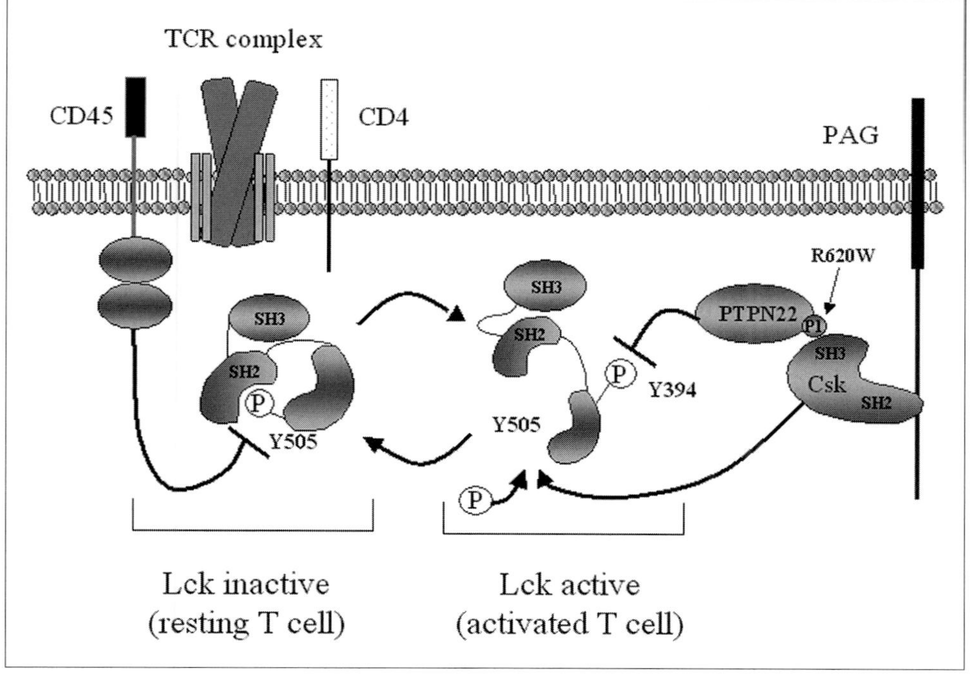

Figure 2

Regulation of Lck activation state. Lck is a src family kinase involved in early T-cell signaling events and is maintained in an inactive state in resting T-cells by phosphorylation of a C-terminal tyrosine 505 (Y505), as depicted at the left side of the figure. Literature suggests that dephosphorylation of Y505 (possibly mediated by CD45) causes a conformational change in Lck, resulting in phosphorylation of an activating tryrosine residue Y394, leading to Lck activation, shown at the right of the figure. PTPN22 binds the tyrosine kinase Csk via a proline-rich SH3 binding site (P1). This binding is thought to enable co-localization of PTPN22 to Lck, dephosphorylation of Lck Y394 and return of Lck to an inactive state, with concomitant rephosphorylation of Y505 by Csk.

The PTPN22 620W is associated with many autoimmune diseases that are characterized by a prominent humoral component

Since the original reports of the association of PTPN22 with type 1 diabetes and rheumatoid arthritis, a number of groups have confirmed these associations [9, 55–61]. The importance of replication in an independent sample collection should be emphasized, as the vast majority of candidate gene association studies fail to

Table 1 - Summary of PTPN22 associations with autoimmune diseases

| Disease | cases | | | controls | | | | | |
First author	TT	TC	CC	TT	TC	CC	OR (allele)	OR (CT vs. CC)	OR (TT vs. CC)
RA									
Begovich	6	119	350	3	78	394	1.65 (1.24–2.20)	1.72 (1.25–2.36)	2.25 (0.56–9.07)
Begovich	21	241	578	9	143	774	2.13 (1.73–2.61)	2.26 (1.79–2.85)	3.12 (1.42–6.87)
Lee et al.	41	287	1085	12	221	1168	1.57 (1.32–1.86)	1.40 (1.15–1.70)	3.68 (1.92–7.04)
Orozco	8	155	663	7	139	890	1.45 (1.15–1.82)	1.50 (1.17–1.92)	1.53 (0.55–4.25)
Hinks	27	262	597	9	105	481	1.88 (1.51–2.35)	2.01 (1.56–2.60)	2.42 (1.15–5.07)
Zhernakova	2	40	109	4	84	440	1.79 (1.22–2.61)	1.92 (1.26–2.94)	2.02 (0.38–10.80)
Steer	12	72	218	1	61	312	2.05 (1.47–2.87)	1.69 (1.15–2.47)	17.17 (3.79–77.80)
Plenge	43	389	1081	16	187	671	1.30 (1.09–1.54)	1.29 (1.06–1.58)	1.67 (0.94–2.97)
						pooled	**1.75 (1.60–1.91)**	**1.72 (1.56–1.91)**	**3.05 (2.15–4.34)**
T1D									
Bottini_N.Am	11	90	193	4	84	307	1.79 (1.33–2.40)	1.70 (1.21–2.41)	4.37 (1.50–12.72)
Bottini_Italian	1	15	158	0	9	205	2.39 (1.08–5.31)	2.16 (0.94–4.98)	n/a
Smyth	40	456	1077	18	323	1377	1.76 (1.52–2.03)	1.81 (1.53–2.13)	2.84 (1.62–4.98)
Zhernakova	12	96	226	4	84	440	2.29 (1.71–3.07)	2.22 (1.60–3.09)	5.84 (1.86–18.32)
Zheng	9	97	290	8	186	984	1.81 (1.42–2.31)	1.77 (1.34–2.33)	3.82 (1.46–9.98)
						pooled	**1.84 (1.66–2.05)**	**1.84 (1.63–2.08)**	**3.54 (2.33–5.37)**
SLE									
Kyogoku	21	136	548	12	315	1634	1.53 (1.26–1.85)	1.29 (1.03–1.61)	5.22 (2.74–9.94)
Orozco	2	62	274	4	63	445	1.45 (1.02–2.06)	1.60 (1.09–2.34)	0.81 (0.15–4.45)
						pooled	**1.51 (1.27–1.79)**	**1.36 (1.12–1.65)**	**3.73 (1.97–7.05)**
Graves'									
Smyth	18	222	661	10	154	669	1.43 (1.17–1.76)	1.46 (1.16–1.84)	1.82 (0.84–3.93)
Skorka	11	90	189	4	68	238	1.71 (1.25–2.35)	1.67 (1.16–2.40)	3.46 (1.16–10.35)
Velaga	6	139	404	3	61	365	1.88 (1.40–2.54)	2.06 (1.48–2.86)	1.81 (0.46–7.14)
						pooled	**1.60 (1.38–1.85)**	**1.64 (1.39–1.95)**	**2.16 (1.21–3.88)**
JIA									
Hinks	16	166	479	9	105	481	1.53 (1.20–1.94)	1.59 (1.21–2.09)	1.79 (0.78–4.08)2

replicate [62]. The association of PTPN22 to disease susceptibility has been extended to other autoimmune diseases such as autoimmune thyroid disease [9, 10, 63] and systemic lupus [8, 61] and juvenile inflammatory arthritis [60]. However, neither multiple sclerosis [64] nor Crohn's disease [65] appear to be associated with the PTPN22 620W allele. In all associated diseases, inheritance fits best under a multiplicative genetic model: 620W/W homozygotes have a substantial increase in relative risk compared to heterozygotes. This observation is consistent with a threshold effect of the risk allele on signaling in T-cells. The results from a variety of case-control studies are summarized in Table 1.

Interestingly, the PTPN22 associated disorders (T1D, RA, SLE, and AITD) are also diseases that tend to cluster together in families, as discussed in the first sections of this review. Indeed, these PTPN22 associations have been replicated directly in such multiplex families [11]. In addition, all of these diseases are accompanied by a prominent humoral component. In some cases these autoantibodies are clearly pathogenic (Grave's disease and some lupus autoantibodies), whereas a direct pathogenic role for autoantibodies is less clear for RA, Hashimotos thyroiditis and T1D. Nevertheless, in all of these disorders, autoantibodies appear months or years prior to the onset of clinical disease. In view of the increased germinal centers and elevated immunoglobulin levels in the PEP knockout mouse [51], it is tempting to speculate that PTPN22 620W allele may predispose to the development of these autoantibodies. However, this issue has not yet been directly addressed. Specifically, there is no information on whether PTPN22 620W is associated with the development of the autoantibodies in the absence of clinical disease, or in contrast, whether PTPN22 might be associated with progression to overt autoimmunity rather than autoantibody development *per se*. These questions will require additional studies in the appropriate human populations.

Wide variation in allele frequency of PTPN22 R620W in different populations: evidence for selection?

The initial reports of disease association were reported in predominantly white European-American populations, where the 1858T (620W) risk allele frequency is approximately 8%. Further investigation revealed that this allele is virtually absent in both East Asian and African populations. We have extended these findings by genotyping the allele in a worldwide diversity panel [66]. Our unpublished results on this panel, as well as in other white European population subgroups [67] are summarized in Table 2. It is interesting to note the dramatic differences in allele frequency across the globe. The 620W allele is extremely rare in both African and East Asian populations. Furthermore, there is a significant north–south cline within Europe; the allele is present in over 12% of Swedish individuals, whereas subjects from southern Europe generally have much lower allele frequencies [6].

Table 2 - PTPN22 R620W allele frequency distribution in various populations

Geographic region/ethnicity	No. individuals	allele freq (95% CI)
Central/South American[1]	108	0.000 (0.000-0.001)
E. Asian[1]	226	0.002 (0.000-0.007)
African[1]	150	0.003 (0.000-0.010)
Middle Eastern[1]	348	0.017 (0.008-0.027)
European[1]	120	0.038 (0.013-0.062)
Melanesian[1]	39	0.051 (0.002-0.100)
Central Asian[1]	67	0.075 (0.030-0.119)
Swedish controls[2]	874	0.125 (0.111-0.142)
Southern European-American[3]	471	0.036 (0.026-0.050)
Northern European-American[3]	261	0.109 (0.085-0.138)

[1]*Human Genome Diversity Panel [66]*
[2]*reference [75]*
[3]*Unpublished data, derived from New York Cancer Project cohort [67]*

Although there are many possible explanations for these observations, including founder effects and random genetic drift, it is also possible that this allele frequency distribution may reflect the effects of positive selective pressures during human history. There are several distinct methods to test for positive selection [68]. All accepted tests require empiric data on genome-wide patterns of genetic diversity, and these datasets are only now becoming available through projects such as the International Haplotype Map (HapMap) Project [69]. No single genetic test can establish definitive evidence for or against positive selection; these tests simply provide an empiric distribution of whether a pattern of genetic diversity is unusual or not. Definitive proof requires functional and biological data.

To further investigate this possibility, we performed one test of selection that examines the extent of LD surrounding a haplotype [70]. With this test of selection, we asked whether the susceptible 620W allele resides on a haplotype that exhibits extensive linkage disequilibrium (LD), beyond that expected based on its population frequency. The principle behind this test is that an allele that has undergone positive selection increased in population frequency faster than meiotic recombination broke down local patterns of LD. This genetic phenomenon is often referred to as a 'selective sweep'. Applying the long-range haplotype test to the 620W allele, we find no striking evidence for selection (Robert Plenge, unpublished data). A more thorough analysis is necessary to assess whether the unusual allele frequency is the result of drift, positive selection, or another evolutionary factor (e.g., balancing selection).

If positive selection has shaped the allele frequency distribution, it is intriguing that a key virulence factor of *Yersinia pestis*, the causative agent of plague, mimics the effects of PTPN22 on Lck in order to disable T-cell responses [71]. Immune defenses against *Y. pestis* are disrupted in both phagocytic cells as well as T- and B-cells by the injection of a plasmid encoded protein tyrosine phosphatase, YopH [72]. In T-cells, the activating phosphate at position 394 of Lck is a major substrate for YopH, and it thereby disables the proximal T-cell receptor signaling apparatus. Perhaps reduced PTPN22 availability/activity in 620W allele carriers provided for marginally increased Lck activity in T-cells during the early stages of infection, enhancing recovery rates, and thus selecting for this allele as the plague pandemic moved northward through Europe in the fourteenth century. Of course, the absence of precise information on the functional effects of the PTPN22 620W allele, many alternative explanations are also possible.

Gene discovery in human autoimmune disease: an engine of hypothesis generation

The example of PTPN22 suggests that further research in human genetics is likely to open up many new avenues of investigation in autoimmune diseases. The era of whole genome association studies is now upon us, a clear example of a 'discovery driven', as opposed to a 'hypothesis driven', approach to biomedical research. As noted above, the identification of PTPN22 as a susceptibility allele for RA was a result of a discovery approach. Future understanding of the mechanism of action of PTPN22 will likely involve additional discovery driven experimentation on genomic and proteomic platforms. Thus, a continuing interaction between hypothesis, exploratory discovery and technology development forms a kind of engine of hypothesis generation, as illustrated in Figure 3.

Human genetics is now in the middle of a remarkable acceleration, largely driven by dramatic advances in the technologies for both producing and analyzing genetic data. Studies involving thousands of subjects and millions of genotypes are now a reality, and this is fast becoming a routinely accepted methodology for approaching genetically complex problems such as autoimmunity. This has made it easier for scientists to confront and actually address a fundamental reality, namely that multiple genetic variants (alleles) with modest effect are likely to underlie most autoimmune diseases. At least some of these variants are common in the population, whereas others may be rare. The relative contribution of common *versus* rare variants to disease susceptibility is not known [73, 74]. The identification of the rare variants, if they exist, is still a very difficult proposition. However, finding the common variants is a now a tractable problem, given the availability of appropriate study populations and sufficient financial resources to carry out comprehensive genetic analyses. Fortunately, the costs of the genotyping component have dropped

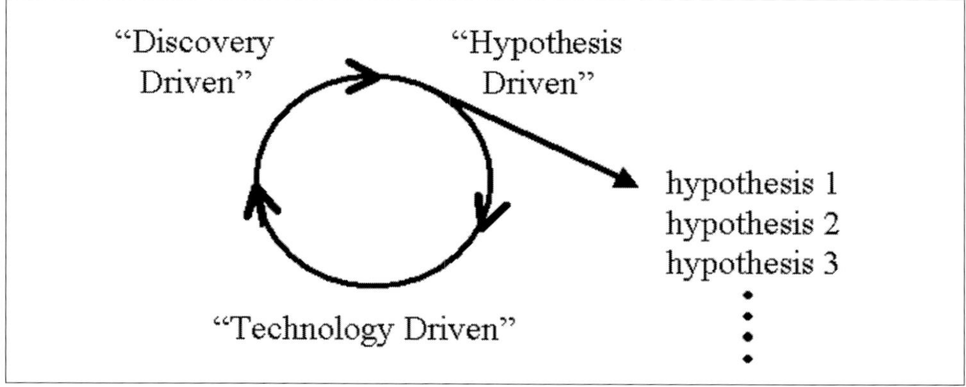

Figure 3
Modern human genetics: an engine of hypothesis generation.

dramatically over the last several years. On the other hand, the costs and complexity of assembling large and well-characterized study populations are likely to remain high and become increasingly important for understanding how the newly discovered genes interact with each other, and with environmental factors, to cause disease. The required resources generally outstrip the capabilities of any single research group or institution. Thus, a collaborative effort among many investigators worldwide is the only rational way to proceed. Fortunately, this is now occurring for many autoimmune diseases, and it is therefore likely that a large fraction of the genes involved in human autoimmunity will be identified by the end of the decade.

Acknowledgements

PKG is supported by grants from the National Arthritis Foundation and the National Institutes of Health. We thank Dr. Franak Batliwalla for critical review and assistance with this manuscript.

References

1   Eisenbarth GS, Gottlieb PA (2004) Autoimmune polyendocrine syndromes. *N Engl J Med* 350: 2068–2079

2   Liston A, Lesage S, Gray DH, Boyd RL, Goodnow CC (2005) Genetic lesions in T-cell tolerance and thresholds for autoimmunity. *Immunol Rev* 204: 87–101

3   Pugliese A, Zeller M, Fernandez A, Jr., Zalcberg LJ, Bartlett RJ, Ricordi C, Pietropaolo M, Eisenbarth GS, Bennett ST, Patel DD (1997) The insulin gene is tran-

scribed in the human thymus and transcription levels correlated with allelic variation at the INS VNTR-IDDM2 susceptibility locus for type 1 diabetes. *Nat Genet* 15: 293–297

4 Sakaguchi S (2005) Naturally arising Foxp3-expressing CD25$^+$CD4$^+$ regulatory T cells in immunological tolerance to self and non-self. *Nat Immunol* 6: 345–352

5 Paust S, Cantor H (2005) Regulatory T cells and autoimmune disease. *Immunol Rev* 204: 195–207

6 Bottini N, Musumeci L, Alonso A, Rahmouni S, Nika K, Rostamkhani M, Mac-Murray J, Meloni GF, Lucarelli P, Pellecchia M et al (2004) A functional variant of lymphoid tyrosine phosphatase is associated with type I diabetes. *Nat Genet* 36: 337–338

7 Begovich AB, Carlton VE, Honigberg LA, Schrodi SJ, Chokkalingam AP, Alexander HC, Ardlie KG, Huang Q, Smith AM, Spoerke JM et al (2004) A missense single-nucleotide polymorphism in a gene encoding a protein tyrosine phosphatase (PTPN22) is associated with rheumatoid arthritis. *Am J Hum Genet* 75: 330–337

8 Kyogoku C, Langefeld CD, Ortmann WA, Lee A, Selby S, Carlton VE, Chang M, Ramos P, Baechler EC, Batliwalla FM et al (2004) Genetic association of the R620W polymorphism of protein tyrosine phosphatase PTPN22 with human SLE. *Am J Hum Genet* 75: 504–507

9 Smyth D, Cooper JD, Collins JE, Heward JM, Franklyn JA, Howson JM, Vella A, Nutland S, Rance HE, Maier L et al (2004) Replication of an association between the lymphoid tyrosine phosphatase locus (LYP/PTPN22) with type 1 diabetes, and evidence for its role as a general autoimmunity locus. *Diabetes* 53: 3020–3023

10 Skorka A, Bednarczuk T, Bar-Andziak E, Nauman J, Ploski R (2005) Lymphoid tyrosine phosphatase (PTPN22/LYP) variant and Graves' disease in a Polish population: association and gene dose-dependent correlation with age of onset. *Clin Endocrinol (Oxf)* 62: 679–682

11 Criswell LA, Pfeiffer KA, Lum RF, Gonzales B, Novitzke J, Kern M, Moser KL, Begovich AB, Carlton VE, Li W et al (2005) Analysis of families in the multiple autoimmune disease genetics consortium (MADGC) collection: the PTPN22 620W allele associates with multiple autoimmune phenotypes. *Am J Hum Genet* 76: 561–571

12 Vyse TJ, Todd JA (1996) Genetic analysis of autoimmune disease. *Cell* 85: 311–318

13 Seldin MF, Amos CI, Ward R, Gregersen PK (1999) The genetics revolution and the assault on rheumatoid arthritis. *Arthritis Rheum* 42: 1071–1079

14 Li CC (1987) A genetical model for emergenesis: in memory of Laurence H. Snyder, 1901-86. *Am J Hum Genet* 41: 517–523

15 Torfs CP, King MC, Huey B, Malmgren J, Grumet FC (1986) Genetic interrelationship between insulin-dependent diabetes mellitus, the autoimmune thyroid diseases, and rheumatoid arthritis. *Am J Hum Genet* 38: 170–187

16 Lin JP, Cash JM, Doyle SZ, Peden S, Kanik K, Amos CI, Bale SJ, Wilder RL (1998) Familial clustering of rheumatoid arthritis with other autoimmune diseases. *Hum Genet* 103: 475–482

17   Barker JM, Yu J, Yu L, Wang J, Miao D, Bao F, Hoffenberg E, Nelson JC, Gottlieb PA, Rewers M et al (2005) Autoantibody "subspecificity" in type 1 diabetes: risk for organ-specific autoimmunity clusters in distinct groups. *Diabetes Care* 28: 850–855

18   Alarcon-Segovia D, Alarcon-Riquelme ME, Cardiel MH, Caeiro F, Massardo L, Villa AR, Pons-Estel BA (2005) Familial aggregation of systemic lupus erythematosus, rheumatoid arthritis, and other autoimmune diseases in 1,177 lupus patients from the GLADEL cohort. *Arthritis Rheum* 52: 1138–1147

19   Taneja V, Singh RR, Malaviya AN, Anand C, Mehra NK (1993) Occurrence of autoimmune diseases and relationship of autoantibody expression with HLA phenotypes in multicase rheumatoid arthritis families. *Scand J Rheumatol* 22: 152–157

20   Ginn LR, Lin JP, Plotz PH, Bale SJ, Wilder RL, Mbauya A, Miller FW (1998) Familial autoimmunity in pedigrees of idiopathic inflammatory myopathy patients suggests common genetic risk factors for many autoimmune diseases. *Arthritis Rheum* 41: 400–405

21   Broadley SA, Deans J, Sawcer SJ, Clayton D, Compston DA (2000) Autoimmune disease in first-degree relatives of patients with multiple sclerosis. A UK survey. *Brain* 123 (Pt 6): 1102–1111

22   McCombe PA, Chalk JB, Pender MP (1990) Familial occurrence of multiple sclerosis with thyroid disease and systemic lupus erythematosus. *J Neurol Sci* 97: 163–171

23   Henderson RD, Bain CJ, Pender MP (2000) The occurrence of autoimmune diseases in patients with multiple sclerosis and their families. *J Clin Neurosci* 7: 434–437

24   Heinlen LD, McClain MT, Kim X, Quintero DR, James JA, Harley JB, Scofield RH (2003) Anti-Ro and anti-nRNP response in unaffected family members of SLE patients. *Lupus* 12: 335–337

25   Scofield RH (1996) Autoimmune thyroid disease in systemic lupus erythematosus and Sjogren' syndrome. *Clin Exp Rheumatol* 14: 321–330

26   Scofield RH (2004) Autoantibodies as predictors of disease. *Lancet* 363: 1544–1546

27   Corporaal S, Bijl M, Kallenberg CG (2002) Familial occurrence of autoimmune diseases and autoantibodies in a Caucasian population of patients with systemic lupus erythematosus. *Clin Rheumatol* 21: 108–113

28   Vianna JL, Haga HJ, Asherson RA, Swana G, Hughes GR (1991) A prospective evaluation of antithyroid antibody prevalence in 100 patients with systemic lupus erythematosus. *J Rheumatol* 18: 1193–1195

29   Schellekens GA, Visser H, de Jong BA, van den Hoogen FH, Hazes JM, Breedveld FC, van Venrooij WJ (2000) The diagnostic properties of rheumatoid arthritis antibodies recognizing a cyclic citrullinated peptide. *Arthritis Rheum* 43: 155–163

30   van Gaalen FA, van Aken J, Huizinga TW, Schreuder GM, Breedveld FC, Zanelli E, van Venrooij WJ, Verweij CL, Toes RE, de Vries RR (2004) Association between HLA class II genes and autoantibodies to cyclic citrullinated peptides (CCPs) influences the severity of rheumatoid arthritis. *Arthritis Rheum* 50: 2113–2121

31   Caron P, Lassoued S, Dromer C, Oksman F, Fournie A (1992) Prevalence of thyroid abnormalities in patients with rheumatoid arthritis. Thyroidology 4: 99–102

32   Silman AJ, Ollier WE, Bubel MA (1989) Autoimmune thyroid disease and thyroid autoantibodies in rheumatoid arthritis patients and their families. Br *J Rheumatol* 28: 18–21

33   Risch N, Merikangas K (1996) The future of genetic studies of complex human diseases. Science 273: 1516–1517

34   Becker KG, Simon RM, Bailey-Wilson JE, Freidlin B, Biddison WE, McFarland HF, Trent JM (1998) Clustering of non-major histocompatibility complex susceptibility candidate loci in human autoimmune diseases. Proc Natl Acad Sci U S A 95: 9979–9984

35   Gaffney PM, Ortmann WA, Selby SA, Shark KB, Ockenden TC, Rohlf KE, Walgrave NL, Boyum WP, Malmgren ML, Miller ME et al (2000) Genome screening in human systemic lupus erythematosus: results from a second Minnesota cohort and combined analyses of 187 sib-pair families. *Am J Hum Genet* 66: 547–556

36   Jawaheer D, Seldin MF, Amos CI, Chen WV, Shigeta R, Etzel C, Damle A, Xiao X, Chen D, Lum RF et al (2003) Screening the genome for rheumatoid arthritis susceptibility genes: a replication study and combined analysis of 512 multicase families. *Arthritis Rheum* 48: 906–916

37   Cox NJ, Wapelhorst B, Morrison VA, Johnson L, Pinchuk L, Spielman RS, Todd JA, Concannon P (2001) Seven regions of the genome show evidence of linkage to type 1 diabetes in a consensus analysis of 767 multiplex families. *Am J Hum Genet* 69: 820–830

38   Ewens KG, Johnson LN, Wapelhorst B, OJ RheumatolBrien K, Gutin S, Morrison VA, Street C, Gregory SG, Spielman RS, Concannon P (2002) Linkage and association with type 1 diabetes on chromosome 1q42. *Diabetes* 51: 3318–3325

39   Graham RR, Langefeld CD, Gaffney PM, Ortmann WA, Selby SA, Baechler EC, Shark KB, Ockenden TC, Rohlf KE, Moser KL et al (2001) Genetic linkage and transmission disequilibrium of marker haplotypes at chromosome 1q41 in human systemic lupus erythematosus. *Arthritis Res* 3: 299–305

40   Gregersen PK (2005) Pathways to gene identification in rheumatoid arthritis: PTPN22 and beyond. *Immunol Rev* 204: 74–86

41   Hardenbol P, Yu F, Belmont J, Mackenzie J, Bruckner C, Brundage T, Boudreau A, Chow S, Eberle J, Erbilgin A et al (2005) Highly multiplexed molecular inversion probe genotyping: over 10,000 targeted SNPs genotyped in a single tube assay. *Genome Res* 15: 269–275

42   Gunderson KL, Steemers FJ, Lee G, Mendoza LG, Chee MS (2005) A genome-wide scalable SNP genotyping assay using microarray technology. *Nat Genet* 37: 549–554

43   Matsuzaki H, Dong S, Loi H, Di X, Liu G, Hubbell E, Law J, Berntsen T, Chadha M, Hui H et al (2004) Genotyping over 100,000 SNPs on a pair of oligonucleotide arrays. *Nat Methods* 1: 109–111

44   Hirschhorn JN, Daly MJ (2005) Genome-wide association studies for common diseases and complex traits. *Nat Rev Genet* 6: 95–108

45 Wang WY, Barratt BJ, Clayton DG, Todd JA (2005) Genome-wide association studies: theoretical and practical concerns. *Nat Rev Genet* 6: 109–118

46 Klein RJ, Zeiss C, Chew EY, Tsai JY, Sackler RS, Haynes C, Henning AK, Sangiovanni JP, Mane SM, Mayne ST et al (2005) Complement factor H polymorphism in age-related macular degeneration. *Science* 308: 385–389

47 Mustelin T, Alonso A, Bottini N, Huynh H, Rahmouni S, Nika K, Louis-dit-Sully C, Tautz L, Togo SH, Bruckner S et al (2004) Protein tyrosine phosphatases in T cell physiology. *Mol Immunol* 41: 687–700

48 Alonso A, Sasin J, Bottini N, Friedberg I, Friedberg I, Osterman A, Godzik A, Hunter T, Dixon J, Mustelin T (2004) Protein tyrosine phosphatases in the human genome. *Cell* 117: 699–711

49 Iivanainen AV, Lindqvist C, Mustelin T, Andersson LC (1990) Phosphotyrosine phosphatases are involved in reversion of T lymphoblastic proliferation. *Eur J Immunol* 20: 2509–2512

50 O'Shea JJ, McVicar DW, Bailey TL, Burns C, Smyth MJ (1992) Activation of human peripheral blood T lymphocytes by pharmacological induction of protein-tyrosine phosphorylation. *Proc Natl Acad Sci USA* 89: 10306–10310

51 Hasegawa K, Martin F, Huang G, Tumas D, Diehl L, Chan AC (2004) PEST domain-enriched tyrosine phosphatase (PEP) regulation of effector/memory T cells. *Science* 303: 685–689

52 Gregorieff A, Cloutier JF, Veillette A (1998) Sequence requirements for association of protein-tyrosine phosphatase PEP with the Src homology 3 domain of inhibitory tyrosine protein kinase p50(csk). *J Biol Chem* 273: 13217–13222

53 Cloutier JF, Veillette A (1999) Cooperative inhibition of T-cell antigen receptor signaling by a complex between a kinase and a phosphatase. *J Exp Med* 189: 111–121

54 Cohen S, Dadi H, Shaoul E, Sharfe N, Roifman CM (1999) Cloning and characterization of a lymphoid-specific, inducible human protein tyrosine phosphatase, Lyp. *Blood* 93: 2013–2024

55 Lee AT, Li W, Liew A, Bombardier C, Weisman M, Massarotti EM, Kent J, Wolfe F, Begovich AB, Gregersen PK (2005) The PTPN22 R620W polymorphism associates with RF positive rheumatoid arthritis in a dose-dependent manner but not with HLA-SE status. *Genes Immun* 6: 129–133

56 Onengut-Gumuscu S, Ewens KG, Spielman RS, Concannon P (2004) A functional polymorphism (1858C/T) in the PTPN22 gene is linked and associated with type I diabetes in multiplex families. *Genes Immun* 5: 678–680

57 Zhernakova A, Eerligh P, Wijmenga C, Barrera P, Roep BO, Koeleman BP (2005) Differential association of the PTPN22 coding variant with autoimmune diseases in a Dutch population. *Genes Immun* 6: 459–461

58 Zheng W, She JX (2005) Genetic association between a lymphoid tyrosine phosphatase (PTPN22) and type 1 diabetes. *Diabetes* 54: 906–908

59 Steer S, Lad B, Grumley JA, Kingsley GH, Fisher SA (2005) Association of R602W in a protein tyrosine phosphatase gene with a high risk of rheumatoid arthritis in a

British population: evidence for an early onset/disease severity effect. *Arthritis Rheum* 52: 358–360

60 Hinks A, Barton A, John S, Bruce I, Hawkins C, Griffiths CE, Donn R, Thomson W, Silman A, Worthington J (2005) Association between the PTPN22 gene and rheumatoid arthritis and juvenile idiopathic arthritis in a UK population: Further support that PTPN22 is an autoimmunity gene. *Arthritis Rheum* 52: 1694–1699

61 Orozco G, Sanchez E, Gonzalez-Gay MA, Lopez-Nevot MA, Torres B, Caliz R, Ortego-Centeno N, Jimenez-Alonso J, Pascual-Salcedo D, Balsa A et al (2005) Association of a functional single-nucleotide polymorphism of PTPN22, encoding lymphoid protein phosphatase, with rheumatoid arthritis and systemic lupus erythematosus. *Arthritis Rheum* 52: 219–224

62 Ioannidis JP, Ntzani EE, Trikalinos TA, Contopoulos-Ioannidis DG (2001) Replication validity of genetic association studies. *Nat Genet* 29: 306–309

63 Velaga MR, Wilson V, Jennings CE, Owen CJ, Herington S, Donaldson PT, Ball SG, James RA, Quinton R, Perros P et al (2004) The codon 620 tryptophan allele of the lymphoid tyrosine phosphatase (LYP) gene is a major determinant of GravesJ Rheumatol disease. *J Clin Endocrinol Metab* 89: 5862–5865

64 Begovich AB, Caillier SJ, Alexander HC, Penko JM, Hauser SL, Barcellos LF, Oksenberg JR (2005) The R620W polymorphism of the protein tyrosine phosphatase PTPN22 is not associated with multiple sclerosis. *Am J Hum Genet* 76: 184–187

65 van Oene M, Wintle RF, Liu X, Yazdanpanah M, Gu X, Newman B, Kwan A, Johnson B, Owen J, Greer W et al (2005) Association of the lymphoid tyrosine phosphatase R620W variant with rheumatoid arthritis, but not Crohn' disease, in Canadian Populations. *Arthritis Rheum* 52: 1993–1998

66 Rosenberg NA, Pritchard JK, Weber JL, Cann HM, Kidd KK, Zhivotovsky LA, Feldman MW (2002) Genetic structure of human populations. *Science* 298: 2381–2385

67 Mitchell MK, Gregersen PK, Johnson S, Parsons R, Vlahov D (2004) The New York Cancer Project: rationale, organization, design, and baseline characteristics. *J Urban Health* 81: 301–310

68 Bamshad M, Wooding SP (2003) Signatures of natural selection in the human genome. *Nat Rev Genet* 4: 99–111

69 The International HapMap Consortium (2003) The International HapMap Project. *Nature* 426: 789–796

70 Sabeti PC, Reich DE, Higgins JM, Levine HZ, Richter DJ, Schaffner SF, Gabriel SB, Platko JV, Patterson NJ, McDonald GJ et al (2002) Detecting recent positive selection in the human genome from haplotype structure. *Nature* 419: 832–837

71 Alonso A, Bottini N, Bruckner S, Rahmouni S, Williams S, Schoenberger SP, Mustelin T (2004) Lck dephosphorylation at Tyr-394 and inhibition of T cell antigen receptor signaling by Yersinia phosphatase YopH. *J Biol Chem* 279: 4922–4928

72 Cornelis GR (2002) Yersinia type III secretion: send in the effectors. *J Cell Biol* 158: 401–408

73   Pritchard JK, Cox NJ (2002) The allelic architecture of human disease genes: common disease-common variant or not? *Hum Mol Genet* 11: 2417–2423

74   Smith DJ, Lusis AJ (2002) The allelic structure of common disease. *Hum Mol Genet* 11: 2455–2461

75   Plenge RM, Padyukov L, Remmers EF, Purcell S, Lee AT, Wolfe F, Kastner DL, Alfredsson L, Altshuler D, Gregersen PK et al (2005) Replication of putative candidate gene associations with rheumatoid arthritis in over 4,000 samples from North America and Sweden: association of susceptibility with PTPN22, CTLA4 and PADI4. *Am J Human Genet* 77: 1044–1060

Shared genes in rheumatic diseases, the role of PD1 and the *RUNX* genes in disease susceptibility

Marta E. Alarcón-Riquelme and Sergey V. Kozyrev

Department of Genetics and Pathology, Rudbeck Laboratory, Uppsala University, Dag Hammarskjölds väg 20, 751 85 Uppsala, Sweden

Introduction

Some rheumatic diseases, such as systemic lupus erythematosus, rheumatoid arthritis, and Sjögren's syndrome, have an autoimmune pathogenesis. Features such as the presence of autoantibodies and immune cell infiltration in tissues are hallmarks of these entities. The hypothesis that pathogenic mechanisms are in common appears to be reasonably justified. Supporting this is the fact that there is familial aggregation of autoimmune clinical manifestations in families of patients with any of these diseases, suggesting that even genetic factors (and environmental factors) are shared. A recent study following 1,177 lupus patients identified strong familial aggregation of rheumatoid arthritis cases in families of lupus probands [1]. The concept of shared autoimmunity was therefore recently put forward, and it has attained support by findings that there is association with variants of genes shared among the diseases. The two most recent examples are *PDCD1* and the *RUNX* genes.

PDCD1

PDCD1 is the gene coding for the immunoreceptor PD-1. PD-1, a 55 kDa protein contains a tyrosine inhibitory motif (ITIM) and a tyrosine switch motif (ITSM) in its intracellular domain, and has been found to be an inhibitor of cellular activation [2–4]. It is thus postulated that PD-1 may regulate peripheral tolerance by attenuating TCR/BCR activation after encounter with suboptimal rather than optimal doses of antigen. Mice made deficient for *pd1* develop a lupus like disease with mild glomerulonephritis and arthritis, and autoimmune dilated cardiomyopathy, depending on the genetic background of the strain [5, 6]. The gene for PD-1, *PDCD1* has been mapped to 2q37.3 [7], a region identified linked to SLE in families of Nordic origin [8, 9]. The analysis of SNPs and haplotypes led to the identification of genetic association with the allele of one polymorphism (named PD1.3A) and SLE [10]. This allele was found to disrupt a binding site for a transcription factor *RUNX1*, as tested using electrophoretic mobility shift and supershift assays. A second study

The Hereditary Basis of Rheumatic Diseases, edited by Rikard Holmdahl
© 2006 Birkhäuser Verlag Basel/Switzerland

identified PD1.3A associated with type 1 diabetes in the Danish population [11], but failed to identify association with PD1.3A and SLE, although a tendency was observed [12]. In contrast, association with SLE nephritis was found with a second polymorphism shown to be also potentially disrupting the binding site of another regulator of transcription, ZEB, however this has not been experimentally shown [12]. A study also performed in Swedish RA patients identified association of PD1.3A with those patients being negative for the rheumatoid factor and the HLA shared epitope [13]. It should be noted that the patients with RA were of recent onset and it is highly feasible that many of these would develop SLE later. Follow up of these individuals would be required. Nevertheless, association has been found for a third polymorphism PD1.5T in Chinese patients with RA [14]. This study has not been replicated in a second set of Chinese RA patients; instead, association was found with a haplotype formed by the polymorphisms PD1.1 located in the promoter of *PDCD1* and the synonymous polymorphism PD1.5 (T-C, Ala-Ala) [15] while PD1.3 was non-polymorphic in this population. It is now becoming clear from these studies that the genetic contribution to susceptibility to rheumatic diseases by *PDCD1* results from various polymorphisms identified for this gene, although PD1.3 is the major polymorphism in Caucasians.

The functional polymorphism PD1.3 is located in the fourth intron of *PDCD1* at a regulatory repeat with binding sites for several transcription factors. As said, the presence of the A allele disrupts the binding site for the transcription factor *RUNX1* [10]. In a way a pathway evolves.

The *RUNX* genes

RUNX1 is one of the three human genes belonging to the family of highly conserved runt-domain-encoding genes [16]. These are *RUNX1* located on human chromosome 21q22.3, *RUNX2* mapped to chromosome 6p21 and *RUNX3* lying on chromosome 1p36 [17–19]. All three regions have been linked to one or several autoimmune or rheumatic diseases [20]. *RUNX* proteins possess a striking feature to either repress or activate transcription of a large variety of genes depending on the co-factors present and architecture of the target sequence [21]. Another feature is a common binding site recognised by all *RUNX* proteins, a fact that hampers their functional analyses [22, 23].

The three *RUNX* genes show extensive structural similarities that include not only a highly conserved region encoding the runt domain, but also the presence of two promoters P1 (distal) and P2 (proximal) and a number of alternative splice isoforms [24–29]. Besides, *RUNX* proteins function as heterodimers formed by interaction with the CBFβ subunit [30], which, first, stabilises the *Runx* proteins by protecting them from rapid ubiquitin-mediated proteolysis [31] and, second, enhances their DNA-binding affinity by several fold [32, 33].

RUNX1 (also known as AML1, PEBP2αB or CBFA2) was primarily implicated in acute myeloid leukaemia. Chromosomal translocations involving AML1 gene (t8;21) were found in cells from 18% of patients with acute myeloid leukaemia (AML), (t12;21) accounting for approximately 25% of ALL paediatric cases, and the most rare rearrangements (t3;21) were associated with CML [34–36]. Although, the question of whether the fusion proteins themselves are the initial cause of the cancer, or it is the consequence of insufficiency of the normal *RUNX1* gene function, or the combination of both, remains open. It was shown that gene dosage is very important for the function of all *RUNX* genes.

RUNX2 (AML3/PEBP2βA/CBFA1) has a major role in bone and cartilage formation by regulating several target genes expressed in these cells [37, 38]. Limb shortening due to disturbances in chondrocyte development, vascular invasion, osteoclast differentiation and periosteal bone formation were observed in a mouse model of dominant negative *Runx2* [39]. Mutations in *RUNX2* lead to a facial craniodysplasia syndrome in humans [40].

RUNX3 (AML2/PEBP2αC/CBFA3) is the shortest of the three genes. About half of the patients with gastric cancer show substantially decreased levels of *Runx3* in epithelial cells from gastric mucosa [41]. Both hemizygous deletions and epigenetic silencing by hypermethylation of the promoter region were found to be involved in it [42]. *RUNX3* appears to be also involved in chondrocyte proliferation and maturation [39].

Thus, in short, the *RUNX* proteins are expressed primarily in skeletal and hematopoietic tissues and cells. In skeletal tissues, their main sites of expression are osteoblasts and chondrocytes, while they are expressed differentially at various stages of hematopoietic development. *RUNX1* has been shown to be a major regulator of haematopoiesis, as mice with dominant negative mutations die early of a lack of haematopoietic development [43]. Similarly, mice deficient for *RUNX2* have important skeletal abnormalities. Mice deficient for *RUNX3* show development of gastric cancer. However, more recent studies have shown an important role of the *RUNX* proteins in immune system development and inflammation.

There is a fine regulation of the *RUNX* genes that is not completely understood or even fully described. As these genes can be expressed concomitantly within the same cell but at different levels, precise transcriptional and translational control of such expression may be crucial for the regulation of target genes. A very elegant model has been recently proposed for the silencing of the CD4 locus during T-cell development towards CD8+ commitment [44, 45]. *RUNX1* was shown to be an active transcriptional repressor of the CD4 gene in double negative (DN) immature thymocytes, while *RUNX3* maintained silencing of CD4 in CD8+SP T-cells through epigenetic mechanisms, showing the context dependency of the regulatory effects of both proteins. *RUNX2* was also found expressed in the thymus, but its role has not been established [46].

All three *RUNX* genes contain multiple recognition sites for themselves in the promoter regions, which makes autoregulation and/or cross-regulation possible. The two examples have been presented so far with RUNX2 autoregulated by negative feedback during osteoblast differentiation and skeletal development [47], and *RUNX3* downregulating *RUNX1* gene in human B-cells [48] by binding to its P1 promoter. Competition between RUNX proteins to exert its functions for association with CBFβ might also contribute to the transcriptional control of target genes [49].

It is now accepted that the *RUNX* proteins are important in TGF-β signal transduction. TGF-β is a pleiotropic cytokine playing a fundamental role in cell growth and differentiation, and known to be one of the strongest immunosuppressor and anti-inflammatory cytokines. Many of the effects of TGF-β are mediated via the *RUNX* proteins [50–53]. This is supported by the demonstration of the physical interaction of the R-Smad transcription factors transmitting the signals from TGF-β receptors into the nucleus [54], and *RUNX*es, and by the finding of functional sites for *RUNX2* in the TGF-β type I receptor promoter [51, 55]. TGF-β induces immunoglobulin class switch recombination in B lymphocytes towards IgA, and this is mediated by *RUNX3* through direct interaction with its binding sites in the Ig Cα promoter. The Smads are also essential for IgA synthesis as disruption of sites for Smads and *RUNX* does impair IgA transcription [56], a further example of the mediation of TGF-β signalling through *RUNX*es [52, 55].

Recently, interesting data has been published as to the role of *RUNX3* in inflammation. Using a KO mouse for *RUNX3*, Fainaru et al. have shown that these mice develop spontaneous eosinophilic lung inflammation and that this is dependent on the signalling through TGF-β [57]. Interestingly, wild type dendritic cells (WT DC) normally overexpress *Runx3* upon activation and respond to TGF-β-mediated inhibition. While KO DC appeared to be insensitive to TGF-β-induced inhibition, accumulated in the alveoli and maintained increased expression of MHC II, OX40L, CD80 and CD86, and thus had the capacity to over-stimulate T lymphocytes, which drive inflammation by enhanced recruitment of eosinophils to the lungs. Furthermore, in the animals developing airway inflammation, there was an increase in TH2 cytokines. At 3 weeks of age, the *RUNX3* KO mice also develop inflammatory bowel disease characterised by leukocyte infiltration, mucosal hyperplasia, lymphoid cluster formation and increased production of IgA [58]. This supports the potential role of the RUNX proteins in inflammatory diseases.

What could the pathogenic role of the *RUNX* genes be in rheumatic diseases? There is still much to be done on this regard. Although, one could anticipate that *RUNX*es are the key transcription factors that may participate in the pathogenesis at two different levels: directly or indirectly. At the first level, their function could be influenced by altered production of diverse proinflammatory/anti-inflammatory stimuli or mediators and thus providing the basis for sustained pathogeneic changes in cell responses or development/maturation or even tissue damages, such as for TGF-β. It is known, for instance that TGF-β production is decreased in lupus nephritis; however it is uncertain if this effect is a cause or consequence of *RUNX* regulation. *RUNX1* and *RUNX3* could be involved in dendritic cell maturation and the subsequent the development of Th2 responses [59], while *RUNX2* could be suspected to be involved as a promoter of joint damage, that is, increasing susceptibility to the development of bone erosions. As *RUNX2* is needed for osteoblast and chondrocyte development, aberrant expression of this regulator may decrease the capacity of the bone to regenerate after damage incurred by the immune system.

Also, mutations of the *RUNX* genes, both in coding and regulatory regions, affecting their function or expression, may change transcriptional patterns of particular cell populations in such a way that could give rise to autoimmunity. It is thus of great interest to analyse if there is an increase in risk of developing of autoimmune diseases in patients with naturally occurring mutations of *RUNX1* gene, such as AML patients with translocations of chromosome 21, individuals with Down syndrome [60, 61] or diseases described caused by monogenic mutations in the *RUNX* genes [62], such as familial mutations leading to haploinsufficiency and a familial platelet disorder [63]. To date, only one report has been made between genetic susceptibility of *RUNX1* and rheumatoid arthritis in Japanese patients [64], however this report analysed only one polymorphism and has not been replicated in other populations. No genetic association with the other *RUNX* genes has been reported.

On another level, *RUNX*es may be involved in the pathogenesis of autoimmune rheumatic diseases indirectly through impaired binding to the target regions affected by regulatory mutations or polymorphisms in the target genes. The examples for the latter case are the PD1.3A polymorphism mentioned above and a SNP disrupting the binding site for *RUNX1* in the organic cation transporter gene SLC22A4 involved in pathogenesis of RA [64] and the example of the loss of the *RUNX* binding site described to be involved in susceptibility for psoriasis [65] in a region between the genes SLC9A3R1 and NAT9, although no particular association has been found with psoriatic arthritis and the described polymorphisms.

Concluding remarks

In this chapter we have reviewed the role of *PDCD1* and the *RUNX* family of genes in the pathogenesis of rheumatic diseases. Much still needs to be done in order to understand how the pathophysiology involving these genes leads to the development of these diseases.

References

1   Alarcón-Segovia D, Alarcón-Riquelme ME, Cardiel MH, Caeiro F, Massardo L, Villa AR, Pons-Estel BA; Grupo Latinoamericano de Estudio del Lupus Eritematoso

(GLADEL) (2005) Familial aggregation of systemic lupus erythematosus, rheumatoid arthritis, and other autoimmune diseases in 1,177 lupus patients from the GLADEL cohort. *Arthritis Rheum* 52: 1138–1147

2   Ishida Y, Agata Y, Shibahara K, Honjo T (1992) Induced expression of PD-1, a novel member of the immunoglobulin gene superfamily, upon programmed cell death. *Embo J* 11: 3887–3895

3   Agata Y, Kawasaki A, Nishimura H, Ishida Y, Tsubata T, Yagita H, Honjo T (1996) Expression of the PD-1 antigen on the surface of stimulated mouse T and B lymphocytes. *Int Immunol* 8: 765–772

4   Nishimura H, Minato N, Nakano T, Honjo T (1998) Immunological studies on PD-1 deficient mice: implication of PD-1 as a negative regulator for B cell responses. *Int Immunol* 10: 1563–1572

5   Nishimura H, Nose M, Hiai H, Minato N, Honjo T (1999) Development of lupus-like autoimmune diseases by disruption of the PD-1 gene encoding an ITIM motif-carrying immunoreceptor. *Immunity* 11: 141–151

6   Nishimura H, Okazaki T, Tanaka Y, Nakatani K, Hara M, Matsumori A, Sasayama S, Mizoguchi A, Hiai H, Minato N, Honjo T (2001) Autoimmune dilated cardiomyopathy in PD-1 receptor-deficient mice. *Science* 291: 319–322

7   Shinohara T, Taniwaki M, Ishida Y, Kawaichi M, Honjo T (1994) Structure and chromosomal localization of the human PD-1 gene (PDCD1). *Genomics* 23: 704–706

8   Lindqvist AK, Steinsson K, Johanneson B, Kristjansdottir H, Arnasson A, Grondal G, Jonasson I, Magnusson V, Sturfelt G, Truedsson L et al (2000) A susceptibility locus for human systemic lupus erythematosus (hSLE1) on chromosome 2q. *J Autoimmun* 14: 169–178

9   Magnusson V, Lindqvist AK, Castillejo-Lopez C, Kristjansdottir H, Steinsson K, Grondal G, Sturfelt G, Truedsson L, Svenungsson E, Lundberg I et al (2000) Fine mapping of the SLEB2 locus involved in susceptibility to systemic lupus erythematosus. *Genomics* 70: 307–314

10 Prokunina L, Castillejo-Lopez C, Oberg F, Gunnarsson I, Berg L, Magnusson V, Brookes AJ, Tentler D, Kristjansdottir H, Grondal G et al (2002) A regulatory polymorphism in PDCD1 is associated with susceptibility to systemic lupus erythematosus in humans. *Nat Genet* 32: 666–669

11 Nielsen C, Hansen D, Husby S, Jacobsen BB, Lillevang ST (2003) Association of a putative regulatory polymorphism in the PD-1 gene with susceptibility to type 1 diabetes. *Tissue Antigens* 62: 492–497

12 Nielsen C, Laustrup H, Voss A, Junker P, Husby S, Lillevang ST (2004) A putative regulatory polymorphism in PD-1 is associated with nephropathy in a population-based cohort of systemic lupus erythematosus patients. *Lupus* 13: 510–516

13 Prokunina L, Padyukov L, Bennet A, de Faire U, Wiman B, Prince J, Alfredsson L, Klareskog L, Alarcon-Riquelme M (2004) Association of the PD-1.3A allele of the PDCD1 gene in patients with rheumatoid arthritis negative for rheumatoid factor and the shared epitope. *Arthritis Rheum* 50: 1770–1773

14 Lin SC, Yen JH, Tsai JJ, Tsai WC, Ou TT, Liu HW, Chen CJ (2004) Association of a programmed death 1 gene polymorphism with the development of rheumatoid arthritis, but not systemic lupus erythematosus. *Arthritis Rheum* 50: 770–775

15 Kong EK, Prokunina-Olsson L, Wong WH, Lau CS, Chan TM, Alarcon-Riquelme M, Lau YL (2005) A new haplotype of PDCD1 is associated with rheumatoid arthritis in Hong Kong Chinese. *Arthritis Rheum* 52: 1058–1062

16 Berardi MJ, Sun C, Zehr M, Abildgaard F, Peng J, Speck NA, Bushweller JH (1999) The Ig fold of the core binding factor alpha Runt domain is a member of a family of structurally and functionally related Ig-fold DNA-binding domains. *Structure Fold Des* 7: 1247–1256

17 Zhang YW, Bae SC, Takahashi E, Ito Y (1997) The cDNA cloning of the transcripts of human PEBP2alphaA/CBFA1 mapped to 6p12.3-p21.1, the locus for cleidocranial dysplasia. *Oncogene* 15: 367–371

18 Zent C, Kim N, Hiebert S, Zhang DE, Tenen DG, Rowley JD, Nucifora G (1996) Rearrangement of the AML1/CBFA2 gene in myeloid leukemia with the 3;21 translocation: expression of co-existing multiple chimeric genes with similar functions as transcriptional repressors, but with opposite tumorigenic properties. *Curr Top Microbiol Immunol* 211: 243–252

19 Bangsow C, Rubins N, Glusman G, Bernstein Y, Negreanu V, Goldenberg D, Lotem J, Ben-Asher E, Lancet D, Levanon D, Groner Y (2001) The RUNX3 gene--sequence, structure and regulated expression. *Gene* 279: 221–232

20 Lindqvist AK, Alarcon-Riquelme ME (1999) The genetics of systemic lupus erythematosus. *Scand J Immunol* 50: 562–571

21 Levanon D, Groner Y (2004) Structure and regulated expression of mammalian RUNX genes. *Oncogene* 23: 4211–4219

22 Ito Y (1999) Molecular basis of tissue-specific gene expression mediated by the runt domain transcription factor PEBP2/CBF. *Genes Cells* 4: 685–696

23 Stein GS, van Wijnen AJ, Stein JL, Lian JB, Pockwinse S, McNeil S (1998) Interrelationships of nuclear structure and transcriptional control: functional consequences of being in the right place at the right time. *J Cell Biochem* 70: 200–212

24 Stewart M, Terry A, Hu M, O'Hara M, Blyth K, Baxter E, Cameron E, Onions DE, Neil JC (1997) Proviral insertions induce the expression of bone-specific isoforms of PEBP2alphaA (CBFA1): evidence for a new myc collaborating oncogene. *Proc Natl Acad Sci USA* 94: 8646–8651

25 Xiao ZS, Thomas R, Hinson TK, Quarles LD (1998) Genomic structure and isoform expression of the mouse, rat and human Cbfa1/Osf2 transcription factor. *Gene* 214: 187–197

26 Xiao ZS, Liu SG, Hinson TK, Quarles LD (2001) Characterization of the upstream mouse Cbfa1/Runx2 promoter. *J Cell Biochem* 82: 647–659

27 Tsuji K, Noda M (2000) Identification and expression of a novel 3'-exon of mouse Runx1/Pebp2alphaB/Cbfa2/AML1 gene. *Biochem Biophys Res Commun* 274: 171–176

28 Stein GS, Lian JB, Stein JL, van Wijnen AJ, Montecino M, Pratap J, Choi J, Zaidi SK,

Javed A, Gutierrez S, Harrington K, Shen J, Young D (2003) Intranuclear organization of RUNX transcriptional regulatory machinery in biological control of skeletogenesis and cancer. *Blood Cells Mol Dis* 30: 170–176

29   Rini D, Calabi F (2001) Identification and comparative analysis of a second runx3 promoter. *Gene* 273: 13–22

30   Backstrom S, Huang SH, Wolf-Watz M, Xie XQ, Hard T, Grundstrom T, Sauer UH (2001) Crystallization and preliminary studies of the DNA-binding runt domain of AML1. *Acta Crystallogr D Biol Crystallogr* 57: 269–271

31   Huang G, Shigesada K, Ito K, Wee HJ, Yokomizo T, Ito Y (2001) Dimerization with PEBP2beta protects RUNX1/AML1 from ubiquitin-proteasome-mediated degradation. *Embo J* 20: 723–733

32   Tahirov TH, Inoue-Bungo T, Morii H, Fujikawa A, Sasaki M, Kimura K, Shiina M, Sato K, Kumasaka T, Yamamoto M, Ishii S, Ogata K (2001) Structural analyses of DNA recognition by the AML1/Runx-1 Runt domain and its allosteric control by CBFbeta. *Cell* 104: 755–767

33   Tahirov TH, Inoue-Bungo T, Sasaki M, Shiina M, Kimura K, Sato K, Kumasaka T, Yamamoto M, Kamiya N, Ogata K (2001) Crystallization and preliminary X-ray analyses of quaternary, ternary and binary protein-DNA complexes with involvement of AML1/Runx-1/CBFalpha Runt domain, CBFbeta and the C/EBPbeta bZip region. *Acta Crystallogr D Biol Crystallogr* 57: 850–853

34   Zent C, Rowley JD, Nucifora G (1997) Rearrangements of the AML1/CBFA2 gene in myeloid leukemia with the 3;21 translocation: *in vitro* and *in vivo* studies. *Leukemia* 11 (Suppl 3): 273–278

35   Lutterbach B, Westendorf JJ, Linggi B, Isaac S, Seto E, Hiebert SW (2000) A mechanism of repression by acute myeloid leukemia-1, the target of multiple chromosomal translocations in acute leukemia. *J Biol Chem* 275: 651–656

36   Lo Coco F, Pisegna S, Diverio D (1997) The AML1 gene: a transcription factor involved in the pathogenesis of myeloid and lymphoid leukemias. *Haematologica* 82: 364–370

37   Komori T, Yagi H, Nomura S, Yamaguchi A, Sasaki K, Deguchi K, Shimizu Y, Bronson RT, Gao YH, Inada M et al (1997) Targeted disruption of Cbfa1 results in a complete lack of bone formation owing to maturational arrest of osteoblasts. *Cell* 89: 755–764

38   Ducy P, Zhang R, Geoffroy V, Ridall AL, Karsenty G (1997) Osf2/Cbfa1: a transcriptional activator of osteoblast differentiation. *Cell* 89: 747–754

39   Stricker S, Fundele R, Vortkamp A, Mundlos S (2002) Role of Runx genes in chondrocyte differentiation. *Dev Biol* 245: 95–108

40   Mundlos S, Otto F, Mundlos C, Mulliken JB, Aylsworth AS, Albright S, Lindhout D, Cole WG, Henn W, Knoll JH et al (1997) Mutations involving the transcription factor CBFA1 cause cleidocranial dysplasia. *Cell* 89: 773–779

41   Li QL, Ito K, Sakakura C, Fukamachi H, Inoue K, Chi XZ, Lee KY, Nomura S, Lee CW, Han SB et al (2002) Causal relationship between the loss of RUNX3 expression and gastric cancer. *Cell* 109: 113–124

42   Li QL, Kim HR, Kim WJ, Choi JK, Lee YH, Kim HM, Li LS, Kim H, Chang J, Ito Y et

al (2004) Transcriptional silencing of the RUNX3 gene by CpG hypermethylation is associated with lung cancer. *Biochem Biophys Res Commun* 314: 223–228

43   Kundu M, Liu PP (2003) Cbf beta is involved in maturation of all lineages of hematopoietic cells during embryogenesis except erythroid. *Blood Cells Mol Dis* 30: 164–169

44   Taniuchi I, Osato M, Egawa T, Sunshine MJ, Bae SC, Komori T, Ito Y, Littman DR (2002) Differential requirements for Runx proteins in CD4 repression and epigenetic silencing during T lymphocyte development. *Cell* 111: 621–633

45   Taniuchi I, Littman DR (2004) Epigenetic gene silencing by Runx proteins. *Oncogene* 23: 4341–4345

46   Vaillant F, Blyth K, Andrew L, Neil JC, Cameron ER (2002) Enforced expression of Runx2 perturbs T cell development at a stage coincident with beta-selection. *J Immunol* 169: 2866–2874

47   Drissi H, Luc Q, Shakoori R, Chuva De Sousa Lopes S, Choi JY, Terry A, Hu M, Jones S, Neil JC, Lian JB et al (2000) Transcriptional autoregulation of the bone related CBFA1/RUNX2 gene. *J Cell Physiol* 184: 341–350

48   Spender LC, Whiteman HJ, Karstegl CE, Farrell PJ (2005) Transcriptional cross-regulation of RUNX1 by RUNX3 in human B cells. *Oncogene* 24: 1873–1881

49   Sakakura C, Hagiwara A, Miyagawa K, Nakashima S, Yoshikawa T, Kin S, Nakase Y, Ito K, Yamagishi H, Yazumi S et al (2005) Frequent downregulation of the runt domain transcription factors RUNX1, RUNX3 and their cofactor CBFB in gastric cancer. *Int J Cancer* 113: 221–228

50   Lee KS, Kim HJ, Li QL, Chi XZ, Ueta C, Komori T, Wozney JM, Kim EG, Choi JY, Ryoo HM, Bae SC (2000) Runx2 is a common target of transforming growth factor beta1 and bone morphogenetic protein 2, and cooperation between Runx2 and Smad5 induces osteoblast-specific gene expression in the pluripotent mesenchymal precursor cell line C2C12. *Mol Cell Biol* 20: 8783–8792

51   Leboy P, Grasso-Knight G, D'Angelo M, Volk SW, Lian JV, Drissi H, Stein GS, Adams SL (2001) Smad-Runx interactions during chondrocyte maturation. *J Bone Joint Surg Am* 83-A (Suppl 1): S15–S22

52   Ito Y, Miyazono K (2003) RUNX transcription factors as key targets of TGF-beta super-family signaling. *Curr Opin Genet Dev* 13: 43–47

53   Tou L, Quibria N, Alexander JM (2003) Transcriptional regulation of the human Runx2/Cbfa1 gene promoter by bone morphogenetic protein-7. *Mol Cell Endocrinol* 205: 121–129

54   Selvamurugan N, Kwok S, Partridge NC (2004) Smad3 interacts with JunB and Cbfa1/Runx2 for transforming growth factor-beta1-stimulated collagenase-3 expression in human breast cancer cells. *J Biol Chem* 279: 27764–27773

55   Ji C, Eickelberg O, McCarthy TL, Centrella M (2001) Control and counter-control of TGF-beta activity through FAST and Runx (CBFa) transcriptional elements in osteoblasts. *Endocrinology* 142: 3873–3879

56   Shi MJ, Park SR, Kim PH, Stavnezer J (2001) Roles of Ets proteins, NF-kappa B and

nocodazole in regulating induction of transcription of mouse germline Ig alpha RNA by transforming growth factor-beta 1. *Int Immunol* 13: 733–767

57 Fainaru O, Woolf E, Lotem J, Yarmus M, Brenner O, Goldenberg D, Negreanu V, Bernstein Y, Levanon D, Jung S, Groner Y (2004) Runx3 regulates mouse TGF-beta-mediated dendritic cell function and its absence results in airway inflammation. *Embo J* 23: 969–979

58 Brenner O, Levanon D, Negreanu V, Golubkov O, Fainaru O, Woolf E, Groner Y (2004) Loss of Runx3 function in leukocytes is associated with spontaneously developed colitis and gastric mucosal hyperplasia. *Proc Natl Acad Sci USA* 101: 16016–16021

59 Komine O, Hayashi K, Natsume W, Watanabe T, Seki Y, Seki N, Yagi R, Sukzuki W, Tamauchi H, Hozumi K et al (2003) The Runx1 transcription factor inhibits the differentiation of naive CD4$^+$ T cells into the Th2 lineage by repressing GATA3 expression. *J Exp Med* 198: 51–61

60 Suwa A, Hirakata M, Satoh S, Ezaki T, Mimori T, Inada S (2000) Systemic lupus erythematosus associated with Down syndrome. *Clin Exp Rheumatol* 18: 650–651

61 Feingold M, Schneller S (1995) Down syndrome and systemic lupus erythematosus. *Clin Genet* 48: 277

62 Osato M (2004) Point mutations in the RUNX1/AML1 gene: another actor in RUNX leukemia. *Oncogene* 23: 4284–4296

63 Song WJ, Sullivan MG, Legare RD, Hutchings S, Tan X, Kufrin D, Ratajczak J, Resende IC, Haworth C, Hock R et al (1999) Haploinsufficiency of CBFA2 causes familial thrombocytopenia with propensity to develop acute myelogenous leukaemia. *Nat Genet* 23: 166–175

64 Tokuhiro S, Yamada R, Chang X, Suzuki A, Kochi Y, Sawada T, Suzuki M, Nagasaki M, Ohtsuki M, Ono M et al (2003) An intronic SNP in a RUNX1 binding site of SLC22A4, encoding an organic cation transporter, is associated with rheumatoid arthritis. *Nat Genet* 35: 341–348. Epub 2003 Nov 9

65 Helms C, Cao L, Krueger JG, Wijsman EM, Chamian F, Gordon D, Heffernan M, Daw JA, Robarge J, Ott J et al (2003) A putative RUNX1 binding site variant between SLC9A3R1 and NAT9 is associated with susceptibility to psoriasis. *Nat Genet* 35: 349–56. Epub 2003 Nov 9

The role of B27 and other genes associated with ankylosing spondylitis

Joachim Sieper and Martin Rudwaleit

Rheumatology, Department of Medicine, Charité, Campus Benjamin Franklin, Berlin, Germany

Introduction

The spondyloarthritides (SpA) comprise ankylosing spondylitis (AS), reactive arthritis (ReA), arthritis/spondylitis with inflammatory bowel disease and arthritis/spondylitis with psoriasis. All these diseases are associated with HLA-B27, although to a different degree (Tab. 1). Further links between these diseases is the possible occurrence of similar clinical symptoms such as inflammatory back pain (predominant in the morning/at night and improvement with exercise) reflecting sacroiliitis/spondylitis, a similar pattern of peripheral joint involvement with an asymmetrical arthritis predominantly of the lower limbs, and in about 20% of patients the manifestations of enthesitis and uveitis. Furthermore, a substantial proportion of the HLA-B27-positive patients with reactive arthritis, arthritis/spondylitis associated with inflammatory bowel disease or psoriasis do develop ankylosing spondylitis (see below). AS is regarded as the SpA with the most severe outcome; it normally starts in the second decade of life. The male to female ratio has more recently been estimated to be around 2:1 [1]. The disease normally starts with inflammation in the sacroiliac joint but can affect the whole spine, with progressive spinal ankylosis as a potential outcome in a subgroup of patients. The association of HLA-B27 with AS/SpA is the highest known MHC Class I association for human diseases and the most relevant single factor for the pathogenesis of SpA. However, other less well defined genetical factors also contribute to the overall genetic risk.

HLA-B27 prevalence in ankylosing spondylitis (AS), background prevalence of HLA-B27 in the population, and the risk to develop AS in HLA-B27 positives

HLA-B27 is positive in 90–95% of patients with ankylosing spondylitis [2] (Tab. 1). There are now considerable epidemiological and transgenic animal model data indi-

The Hereditary Basis of Rheumatic Diseases, edited by Rikard Holmdahl
© 2006 Birkhäuser Verlag Basel/Switzerland

Table 1 - Association of HLA-B27 with spondyloarthritides

	Population of Western/Central European countries	Population of Scandinavian countries	Ankylosing spondylitis	Sacroilitis/ spondylitis with inflammatory bowel disease or psoriasis	Reactive arthritis
HLA-B27 + (%)	6–9%	11–14%	90–95%	50–70%	30–60%

cating a direct role for HLA-B27 in disease pathogenesis, rather than a closely linked gene, but final proof for this is still lacking in humans [3]. It is also clear now that one copy of HLA-B27 (heterozygosity) is sufficient to get the disease. Recent studies of families and twins affected by AS suggest a polygenic model of genetic susceptibility [4]. The concordance rates for monozygotic twins are about 63%, for dizygotic twins about 12.5% and for B27-positive dizygotic twins about 27%. The susceptibility to AS has been estimated to be >90% genetically determined and it has been suggested that there might be, because of this, a rather ubiquitous environmental factor [3, 5].

Most striking is the direct relationship between the prevalence of SpA and the prevalence of HLA-B27 in the general population. Thus, populations with a higher HLA-B27 prevalence have a higher prevalence of SpA and AS. HLA-B27 is present throughout Eurasia (2–15%), but it is absent among genetically unmixed populations of natives in South America, Australia and of equatorial and southern Africa [6]. However, HLA-B27 has a high prevalence (25–50%) among the native peoples of the circumpolar arctic and subarctic regions of Eurasia and North America. The best investigations about the prevalence of SpA are available for AS. Its prevalence has been estimated between 0.2–1.4% in the western world: 0.2% in Holland [7] (HLA-B27 background 8%; AS in HLA-B27-positives: 2%); 0.86% in Germany [8] (HLA-B27 background 9%; AS in HLA-B27-positives 6.4%), and 1.4% in Norway [9] (HLA-B27 background 14%; AS in HLA-B27 positives 6.7%).

However, there are also exceptions from this close correlation demonstrating that some environmental or genetical factors are different in some places of the world. For example, Japan and Lebanon both have a low prevalence of HLA-B27 of 1% and 2–3%, respectively, but 90% of AS patients are HLA-B27-positive in Japan but only 28% in Lebanon. This means that there must be a strong genetical or environmental difference leading to AS in B27-negatives in Lebanon but not in Japan or in other parts of the world. Furthermore, in West Africa (the Gambia), AS is rare despite the presence of B27 (2–3%) in the population [6].

HLA-B27 subtypes and their association with ankylosing spondylitis

HLA-B27 as a serological specificity encompasses 25 different alleles that encode 23 different products (proteins) that have been given the designations B*2701 to B*2723 [10]. These subtypes differ only by small changes, mostly in exons 2 and 3, which encode the alpha 1 and 2 domains of the B27 molecule, respectively. Associations with AS/SpA have until now only been investigated for the first 10 subtypes. The most widespread subtype in the world is B*2705, and epidemiological studies show that the common subtypes B*2705, B*2702, and B*2704 are clearly associated with SpA, while some of the rare or not so well studied subtypes (B*2701, B*2703, B*2707, and B*2708) have also been observed in at least one patient with AS. B*2705 is the predominant subtype in Europeans. The prevalence of B*2702 increases from 10% in northern European B27-positive population, to approximately 20% in Spain, while among Semitic populations (Arabs and Jews), B*2702 prevalence reaches 55%. In northern India the predominant subtypes are B*2705 (50%) and B*2704 (40%), while in West African populations B*2705 (65%) and B*2703 (35%) are the dominant subtypes. B*2704 is the predominant subtype among Chinese and Japanese, and B*2706 in Southeast Asia.

Most interestingly and possibly relevant for pathogenesis, studies from Indonesia, Singapore, Thailand, and Taiwan have reported lack of association of B*2706 with AS and it has also been reported that B*2709, a rare subtype present among Italians, primarily among those residing on the island of Sardinia, is apparently also not associated with AS. B*2709 differs from the disease-associated B*2705 by only one amino acid substitution by the exchange of an Asp^{116} to His^{116} at the peptide binding groove of the HLA-B27-molecule [11] while B*2706 (not disease-associated) differs by only two amino acid substitutions from the disease-associated B*2704 by exchange of His^{114} to Asp^{114} and of Asp^{116} to Tyr^{116} [12].

HLA-B27 and bacteria in the pathogenesis of AS

If HLA-B27 is necessary in the great majority of patients to get the disease, which other factors are needed? There is evidence that bacterial infections or bacterial exposure is an important, if not essential, trigger. First of all, this comes from the relationship between AS with reactive arthritis (ReA) and with inflammatory bowel disease. ReA can occur shortly after a preceding bacterial infection of the gut with enterobacteriae or the urogenital tract with *Chlamydia trachomatis*. 20–40% of these patients develop the full clinical picture of AS 10–20 years after the initial infection, if they are positive for HLA-B27 [13]. Such a relationship between bacterial infections and AS is also supported by older indirect data. In the 1940s and 1950s arthritis associated with urogenital infections was associated with the development of ankylosing spondylitis in Sweden in a higher percentage than 20 to 30 years later [14]. Only consequent and long-term antibiotic treatment later on was the main difference in the management of these patients. Similarly, AS and related diseases were more severe in lower social classes with less hygiene (without a refrigerator) when investigated in North Africa [15], and AS starts at an earlier age and runs a more severe course in countries such as Mexico [16], China [17], and North Africa [18] when compared with Western Europe, suggesting that frequent and repeated bacterial infections play an important role for first disease manifestation at young age and for severity of disease.

Although only less than 10% of AS patients recall an earlier reactive arthritis, many of the gut or urogenital infections preceding the clinical manifestation of ReA can be asymptomatic. Especially infections with *Chlamydia trachomatis* or *Yersinia enterocolitica* go often along with no or only a few clinical symptoms. Indeed *Chlamydia trachomatis* has been detected in synovial fluid or synovial membrane of patients with so-called undifferentiated oligoarthritis (without clear clinical or laboratory evidence of a preceding infection) in up to 30% [19].

Despite the fact that IBD cannot be seen as a bacterial infection stimulation of the immune system by local gut bacteria as a consequence of lesions in the gut mucosa serves most probably a similar purpose. Patients with IBD who are positive for HLA-B27 have an especially high risk to develop AS. In one large study 54% HLA-B27-positive Crohn's patients developed AS, while AS could be diagnosed only in 2.6% of HLA-B27-negative Crohn's patients [20]. Although again less than 10% of patients with primary AS do have clinical evidence of IBD it has become clear that in up to 50% of patients with so-called idiopathic AS macroscopic or microscopic mucosal chronic lesions resembling Crohn's disease can be detected in the gut by colonoscopy [21]. Thus, probably in most, if not all, of the AS patients a bacterial trigger is essential in the pathogenesis of the disease. These patients normally have to be positive for HLA-B27.

This is further supported by the B27-transgenic rat model for AS/SpA. HLA-B27 transgenic rats develop features of SpA including gut inflammation, peripheral arthritis, and psoriasiform skin and nail changes. The importance of environmental factors in this animal model is emphasised by the observation that many of these features, including gut inflammation and arthritis, do not develop in HLA-B27 transgenic rats born and bred in a germ-free environment [22]. Germ-free animals rapidly develop inflammatory disease on removal from the sterile environment. This can be partially prevented by treatment with antibiotics [23].

Models to explain the role of HLA-B27 in the pathogenesis of SpA

Although there is good evidence for the crucial role of HLA-B27 in the pathogenesis of AS and also for an additional role for bacteria it is far less clear how these factors interact. Several models have been proposed, none of which has been proven so far.

Arthritogenic peptide hypothesis

Because the main function of HLA Class I molecules is to present peptide antigens to cytotoxic CD8[+] T-cells it has been proposed that the antigen presenting properties of HLA-B27 could be crucial in the pathogenesis of spondyloarthropathies, the so called 'arthritogenic peptide hypothesis' [24]. It suggests that some B27 subtypes, due to their unique amino acid residues, bind specific arthritogenic peptides, which are recognised by CD8[+] T-cells. Furthermore, in response to these bacterial peptides, autoreactive T-cells recognising antigens with sufficient structural similarity between bacteria and self might become activated by self-peptides.

One major support for this hypothesis comes from the differential association of some of the HLA-B27 subtypes with AS (see above), because the relevance of the single AA-substitutions in the antigen-binding groove of the HLA-B27 molecule can best be explained by presentation of peptide(s) by the susceptible HLA-B27 subtype, but not by the resistant one, to CD8[+] T-cells.

To prove this hypothesis researchers have tried in the past to identify antigens presented by HLA-B27 to CD8[+] T-cells. Most studies were performed first in patients with ReA because in this disease the triggering bacteria are known. After an early report about an HLA-B27-restricted CD8[+] T-cell response against whole bacteria in ReA patients [25] we could demonstrate more recently a synovial CD8[+] T-cell response to a peptide from *Yersinia* heat shock protein 60 in patients with *Yersinia*-induced ReA [26] and also an HLA-B27-restricted CD8[+] T-cell response to peptides derived from several Chlamydial proteins in patients with *Chlamydia*-induced ReA [27]. In the latter study a novel approach for the search of the whole Chlamydial proteom for arthritogenic peptides was applied to identify peptides which stimulate patients' derived CD8[+] T-cells in an HLA-B27 restricted manner. For this we combined two computer prediction algorithms, first, for the binding of peptides to the HLA-B27 molecule and, second, for the cleavage of proteins/peptides by proteasomes in the cytosol [27].

Although a bacterial trigger seems to be necessary to set the disease off there is currently no evidence that microbial antigens do persist in such diverse structures as the sacroliliac joint, the enthesis, or the eye. Thus, CD8[+] T-cell responses directed against autoantigens, which are presented by HLA-B27, have also been investigated in the past. Along this line, a CD8[+] T-cell response to an EBV-epitope derived from the LMP2 protein and to a sequence-related self peptide from the autoantigen 'vasoactive intestinal peptide receptor (VIP) 1' was reported [28], suggesting that molecular mimicry might play a role.

However, such a ubiquitous autoantigen could not explain the location of inflammation (sacroiliac joint and spine) found in AS. More recently, the question about the primary target of the immune response in AS has been discussed intensively [29]. Based mostly on magnetic resonance imaging (MRI) and immunohistology studies several recent articles have proposed that the cartilage is the primary target of the immune response in AS/SpA [29–32]. Therefore, extracellular matrix proteins from human cartilage and fibrocartilage are attractive candidate targets of an immune response in this disorder. The G1-domain of the proteoglycan aggrecan has been of special interest in the past because inflammation could be induced in an animal model resembling AS by immunisation with this protein [33]. We could also show an antigen-specific HLA-B27 restricted CD8[+] T-cell response and the occurrence of arthritis in HLA-B27-transgenic mice immunised with aggrecan-derived peptides [34, 35]. Furthermore, we showed recently that PB CD4[+] T-cells and CD8[+] T-cells derived from AS patients are reactive to aggrecan-specific antigens [36]. However, we could also show that in humans such a CD8[+] T-cell response to peptides derived from the G1-domain of aggrecan is not HLA-B27 restricted [37], making it unlikely that the aggrecan protein is the source of an arthritogenic peptide. Collagens, especially collagen type II, are also important proteins of cartilage. In an earlier study the HLA-B27 restricted CD8[+] T-cell response to four peptides (two from collagen II, two from collagen XI), which showed a good binding to HLA-B27 *in vitro*, were studied in patients with AS and reactive arthritis, but only one ReA patient and none of the AS-patients showed an HLA-B27-restricted CD8[+] T-cell response specific for one of these peptides [38].

Recently we used a more comprehensive approach to look for potentially arthritogenic peptides in AS patients. We investigated all cartilage-derived proteins for the presence of peptides which would be presented by HLA-B27 and would be immunodominant for CD8[+] T-cells derived from the synovial fluid from AS patients. For this we used again the two computer prediction programs (for HLA-B27-binding of peptides and for cleavage of peptides by proteasomes), as described above. Interestingly, we could determine one nonameric peptide from collagen type VI which was recognised by CD8[+] T-cells in four out of seven AS-patients [39]. Further studies are in progress to test whether CD8[+] T-cells specific for cartilage-derived peptide can be found at the site of inflammation.

An oligoclonal expansion of T-cells has also been demonstrated for CD8[+] T-cells in ReA. The synovial fluid derived from different HLA-B27 positive patients suffering from ReA and triggered by different bacteria revealed an astonishing high homology of T-cell receptors [40]. These results lead to the suggestion that similar antigens are recognised by these oligoclonally expanded CD8[+] T-cells which implies the possibility that under certain conditions specific arthritogenic peptides might indeed be produced and presented to the hosts immune system.

Because none of these HLA-B27-restricted CD8[+] T-cell responses were directed against bacterial and/or self antigens, a pathogenetic role could not be proven – other hypotheses have been put forward and tested to explain the association between HLA-B27 and AS and other SpA.

The HLA-B27 misfolding hypothesis and the formation of beta-2-m-free HLA-B27 homodimers

The 'HLA-B27 misfolding hypothesis' states that HLA-B27 itself is directly involved in the pathological process of SpA, a hypothesis that does not include the physiological role of HLA-B27 to present specific antigens. It has been shown that HLA-B27 has a tendency to misfold in the endoplasmatic reticulum which could then induce a proinflammatory stress response [41]. The misfolding is suggested to be due to a particular feature of HLA-B27: newly synthesised HLA-B*2705 seems to fold by and associate with beta 2-microglobuline more slowly in comparison to other MHC Class I molecules.

Allen and co-workers [42] reported that, possibly as a consequence of HLA-B27 misfolding, free HLA-B27 heavy chains (HC) can form abnormal HC homodimers. These researchers could indeed show subsequently that beta(2)m-free HLA-B27 homodimers and multimers are expressed both intracellularly and at the cell surface of leukocytes, dendritic and other cells. They could further demonstrate that such HLA-B27 homodimers expressed by antigen presenting cells can interact with paired immunoglobulin-like receptors on monocytes or B-cells, a process which could then induce or perpetuate immunopathology, independently from specific antigens [43].

HLA-B27-restricted CD4$^+$ T-cell responses

Recently HLA-B27 restricted CD4$^+$ T-cells have been described [44]. It was speculated that the B27-homodimers might mimic MHC Class II molecules and might therefore be capable to present a yet unidentified (auto-)antigen to CD4$^+$ T-cells, which could then be responsible for the local immunopathology found in AS and other SpA [45].

HLA-B27 and intracellular survival of bacteria

A completely different hypothesis is based on the finding that bacteria survive longer after invasion of HLA-B27$^+$ cells compared to HLA-B27-negative cells. Again, the significantly slower folding rate (misfolding) of HLA-B27 intracellularly had been discussed as a possible explanation for this finding [46]. A more recent study could also show in *in vitro* experiments that wild-type HLA-B27-positive cells were more permissive of intracellular replication of *Salmonella enteritides* compared with HLA-A2-transfected control cells [47]. By substituting single amino acids these authors provided evidence that the bacteria permissive phenotype was dependent on glutamic acid substitution at position 45 in the B pocket of the HLA-B27 molecule.

However, the exact mechanism of how this could contribute to a different behaviour of intracellular bacteria was not clarified.

Unfortunately, until now none of the hypotheses presented here to explain the strong association between HLA-B27 and ankylosing spondylitis could be proven. Therefore, more than 30 years after the first description of this association there remains an interesting field for future research.

Other genes associated with ankylosing spondylitis

Twin studies suggest a contribution of HLA-B27 of only 16% of the total genetic risk in AS [48]. Besides HLA-B27 other MHC genes such as HLA-B60 and HLA-DR1 seem to be associated with AS but are of minor importance. The TNF alpha gene is another candidate gene located within the MHC but the role of TNF gene polymorphisms in AS is still controversial [49]. Although MHC is the major gene locus involved in the genetic susceptibility to AS the contribution of the whole MHC locus has been estimated not to exceed 36% of the overall genetic risk [5].

Genome-wide linkage screens performed in the UK and the US have revealed suggestive markers of linkage on chromosomes 1, 2, 5, 6, 9, 10, 16 and 19 in one screen [50] and markers of interest on chromosomes 1, 3, 4, 5, 6, 10, 11, 16, 17, and 19 in the other screen [51]. The discrepancies between the two screens is most likely explained by the small contribution (λs <2.0) of each non-MHC gene to the overall susceptibility to AS and the lack of power in the screens to discern such small effect genes. Some of the suggestive gene markers may include genes associated with diseases that predispose to SpA, such as psoriasis and inflammatory bowel disease, or some may encompass genes relevant for the immune response such as antigen processing and presentation, or the cytokine response. For example, the occurrence of acute anterior uveitis seems to be associated with a gene region located on chromosome 9 [52]. Given the sex bias of AS the X chromosome had also been suspected to be a candidate gene. However, no linkage with AS has been found [53].

Given the central role for bacteria in the pathogenesis of AS and other SpA it has been investigated whether specific cytokine patterns can be found in patients which would allow for bacterial persistence in patients. We could show in ReA patients that expression of T-helper 1 cytokines such as TNF-α and IFN-γ are relatively downregulated and counter-regulatory cytokines such as IL-10 are relatively upregulated and thus might contribute to bacterial persistence, possibly by downregulation of the Th1-cytokines IFN-γ and TNF-α [54–56]. On this background genetical polymorphisms have been investigated for TNF-α and IL-10.

Investigating TNF-α-microsatellites, an association of ReA with a TNFa6-allele has been described; this allele has been associated with a low TNF-α secretion before. Since in this study from Finland TNFa6 was also associated with HLA-B27, the association of TNFa6 with ReA was thought to be secondary to B27 [57]. Two promoter polymorphisms of the TNF-α gene at positions −308 (308.1 and 308.2) and −238 (238.1 and 238.2) have been investigated in AS. The 308.2 genotype was found significantly less frequently in AS than in controls [58]. In some studies 308.2 was associated with higher transcriptional activity. Thus, there is some evidence that TNF-α genotypes which seem to be associated with a low TNF-α-production are present in a higher percentage in ReA- and AS-patients, although available data on this issue are not conclusive [49].

Colleagues from Finland and Germany looked also for IL-10 gene polymorphism in ReA patients. They found a significant decrease in the promoter alleles G12and G10 in the ReA group compared with the HLA-B27-positive controls indicating that these alleles might have a protective effect for the occurrence of ReA [59]. Although it is not clear at the moment whether these alleles are associated with a higher production of IL-10 these data raise the possibility that the relative increase of IL-10 found in ReA might be, at least partially, genetically determined. Studies of IL-10 gene polymorphisms in familial AS revealed no effect on susceptibility to AS but a possible minor role in determining age at onset and disease severity [60].

Transforming growth factor (TGF) beta plays a role in balancing immune responses, bone remodelling and fibrosis, and is therefore one of the candidate cytokines in AS. The gene encoding TGF beta is located on chromosome 19. Increased serum levels of TGF beta 1 and increased production of TGF beta 1 by peripheral blood mononuclear cells upon stimulation with PHA in AS patients from Singapore has recently been reported [61]. Although in this small study there was no association with TGF beta gene polymorphisms a larger study of British and Finnish families suggested at most a small role of the TGF beta gene in AS [62].

The interleukin 1 gene cluster located on chromosome 2 has been found to be significantly involved in AS, as this has been reported by several groups [63–66]. However, which of the many genes located at the IL-1 gene cluster is causatively involved is as yet unclear since the findings of the various studies were disparate. While two studies found significant associations for the gene IL-1RN encoding for the IL-1 receptor antagonist [63, 64], these genetic variants were not confirmed in other studies [65–68], but in two of these studies there was evidence for other genes associated with AS within the IL-1 gene cluster [65, 66].

NOD 2 (CARD15) genotypes which are located on chromosome 16 have been found to be significantly associated with Crohn's disease. NOD2 is expressed in intestinal epithelial cells and may serve as a key component of innate mucosal responses to luminal bacteria as an antibacterial factor. While in a large study no association between the NOD2 variants Pro269Ser, Arg702Trp, Gly908Arg, Leu1007fsinsC and primary AS was found, an association was identified between Gly908Arg and ulcerative spondyloarthritis and an inverse association between Pro268Ser and ulcerative spondyloarthritis with only non-significant trends for spondyloarthritis associated with Crohn's disease [69]. The frequency of these variants was higher in patients with Crohn's disease with radiographic sacroiliitis compared to Crohn's disease without radiographic sacroiliitis [70]. The major CARD15 polymorphisms were found not to be associated with primary AS in two other studies [71, 72]. Overall, primary AS seems not be associated with NOD2/CARD15 polymorphisms but these variants may have a role in the pathophysiological interplay between Crohn's disease and radiographic sacroiliitis/AS.

Other candidate genes which have been investigated are the cytochrome P450 complex with a small effect of homozygosity of the poor metaboliser variant CYP2D6*4 in AS [73] and a possible effect of CYP1A1 was reported in a small case-control study only [74]. No effects have been reported for gene polymorphisms of the Toll-like receptor 4 [75] and of matrix metalloproteinase 3 [76] in recent studies.

Normal osteogenesis and bone formation depends on the level of inorganic pyrophosphate (PPi). The export of PPi from the cell is regulated by the ANK gene, and a mutant of ANK leads to bony ankylosis in mice. Thus, genetic variants of ANKH have been studied in AS and revealed disparate results. While there was neither linkage nor association in the study of more than 200 families from the UK [77], the study from the US on a similar number of families found two ANKH haplotypes to be associated with AS in a gender-specific fashion, one of which associated with male AS and the other one with female AS [78]. Again, the gene effect was small which may account for inconsistent results between the two studies. Of course, other factors such as intrinsic differences between the two populations could also explain the disparate results.

In summary, apart from the MHC gene locus with HLA-B27 itself as probably the most important gene there is clear evidence of contribution of other genes to the susceptibility to AS. However, the contribution of other genes is likely to be small rendering the identification of these genes into a difficult task. Among the various candidate genes investigated so far, the IL-1 gene locus is likely to play a role.

References

1   Braun J, Sieper J (2003) Spondyloarthritides and related arthritides. In: Warrel TM Cox, JD Firth, EJ Benz (eds): *Oxford Textbook of Medicine*, 4th Edition, Oxford University Press, New York 43–53

2   Brewerton DA, Hart FD, Nicholls A, Caffrey M, James DC, Sturrock RD (1973) Ankylosing spondylitis and HL-A 27. *Lancet* 28: 1(7809): 904–907

3   Brown MA, Crane AM, Wordsworth BP (2002) Genetic aspects of susceptibility, severity, and clinical expression in ankylosing spondylitis. *Curr Opin Rheumatol* 14: 354–360

4   Brown MA, Laval SH, Brophy S, Calin A (2000) Recurrence risk modelling of the genet-

ic susceptibility to ankylosing spondylitis. *Ann Rheum Dis* 59: 883–886

5   Brown MA, Kennedy LG, MacGregor AJ, Darke C, Duncan E, Shatford JL, Taylor A, Calin A, Wordsworth P (1997) Susceptibility to ankylosing spondylitis in twins: the role of genes, HLA, and the environment. *Arthritis Rheum* 40: 1823–1828

6   Khan MA (1998) Spondyloarthropathies. *Curr Opin Rheumatol* 10: 279–281

7   van der Linden SM, Valkenburg HA, de Jongh BM, Cats A (1984) The risk of developing ankylosing spondylitis in HLA-B27 positive individuals. A comparison of relatives of spondylitis patients with the general population. *Arthritis Rheum* 27: 241–249

8   Braun J, Bollow M, Remlinger G, Eggens U, Rudwaleit M, Distler A, Sieper J (1998) Prevalence of spondylarthropathies in HLA B27-positive and -negative blood donors. *Arthritis Rheum* 41: 58–67

9   Johnsen K, Gran JT, Dale K, Husby G (1992) The prevalence of ankylosing spondylitis among Norwegian Samis (Lapps). *J Rheumatol* 19: 1591–1594

10   Ball EJ, Khan MA (2001) HLA-B27 polymorphism. Joint Bone Spine 68: 378–382

11   Fiorillo MT, Greco G, Maragno M, Potolicchio I, Monizio A, Dupuis ML, Sorrentino R (1998) The naturally occurring polymorphism Asp116-->His116, differentiating the ankylosing spondylitis-associated HLA-B*2705 from the non-associated HLA-B*2709 subtype, influences peptide-specific CD8 T cell recognition. *Eur J Immunol* 28: 2508–2516

12   D'Amato M, Fiorillo MT, Carcassi C, Mathieu A, Zuccarelli A, Bitti PP, Tosi R, Sorrentino R (1995) Relevance of residue 116 of HLA-B27 in determining susceptibility to ankylosing spondylitis. *Eur J Immunol* 25: 3199–3201

13   Leirisalo-Repo M (1998) Prognosis, course of disease, and treatment of the spondyloarthropathies. *Rheum Dis Clin North Am* 24: 737–751, viii

14   Olhagen B (1983) Urogenital syndromes and spondarthritis. *Br J Rheumatol* 22 (4 Suppl 2): 33–40

15   Claudepierre P, Gueguen A, Ladjouze A, Hajjaj-Hassouni N, Sellami S, Amor B, Dougados M (1995) Predictive factors of severity of spondyloarthropathy in North Africa. Br *J Rheumatol* 34: 1139–1145

16   Lau CS, Burgos-Vargas R, Louthrenoo W, Mok MY, Wordsworth P, Zeng QY (1998) Features of spondyloarthritis around the world. *Rheum Dis Clin North Am* 24: 753–770

17   Huang F, Zhang J, Zhu J, Guo J, Yang C (2003) Juvenile spondyloarthropathies: the Chinese experience. *Rheum Dis Clin North Am* 29: 531–547

18   Hajjaj-Hassouni N, Maetzel A, Dougados M, Amor B (1993) Comparison of patients evaluated for spondylarthropathy in France and Morocco. *Rev Rhum Ed Fr* 60: 420–425

19   Sieper J, Braun J, Kingsley GH (2000) Report on the Fourth International Workshop on Reactive Arthritis. *Arthritis Rheum* 43: 720–734

20   Purrmann J, Zeidler H, Bertrams J, Juli E, Cleveland S, Berges W, Gemsa R, Specker C,

Reis HE (1988) HLA antigens in ankylosing spondylitis associated with Crohn's disease. Increased frequency of the HLA phenotype B27,B44. *J Rheumatol* 15: 1658–1661

21 Mielants H, Veys EM, Cuvelier C, de Vos M (1988) Ileocolonoscopic findings in seronegative spondylarthropathies. *Br J Rheumatol* 27 (Suppl 2): 95–105

22 Taurog JD, Richardson JA, Croft JT, Simmons WA, Zhou M, Fernandez-Sueiro JL, Balish E, Hammer RE (1994) The germfree state prevents development of gut and joint inflammatory disease in HLA-B27 transgenic rats. *J Exp Med* 180: 2359–2364

23 Rath HC, Schultz M, Freitag R, Dieleman LA, Li F, Linde HJ, Scholmerich J, Sartor RB (2001) Different subsets of enteric bacteria induce and perpetuate experimental colitis in rats and mice. *Infect Immun* 69: 2277–2285

24 Kuon W, Sieper J (2003) Identification of HLA-B27-restricted peptides in reactive arthritis and other spondyloarthropathies: computer algorithms and fluorescent activated cell sorting analysis as tools for hunting of HLA-B27-restricted chlamydial and autologous crossreactive peptides involved in reactive arthritis and ankylosing spondylitis. *Rheum Dis Clin North Am* 29: 595–611

25 Hermann E, Yu DT, Meyer zum Buschenfelde KH, Fleischer B (1993) HLA-B27-restricted CD8 T cells derived from synovial fluids of patients with reactive arthritis and ankylosing spondylitis. *Lancet* 342: 646–650

26 Ugrinovic S, Mertz A, Wu P, Braun J, Sieper J (1997) A single nonamer from the Yersinia 60-kDa heat shock protein is the target of HLA-B27-restricted CTL response in *Yersinia*-induced reactive arthritis. *J Immunol* 159: 5715–5723

27 Kuon W, Holzhütter HG, Appel H (2001) Identification of HLA-B27-restricted peptides from the *Chlamydia trachomatis* proteome with possible relevance to HLA-B27-associated diseases. *J Immunol* 167: 4738–4746

28 Fiorillo MT, Maragno M, Butler R, Dupuis ML, Sorrentino R (2000) CD8+ T-cell autoreactivity to an HLA-B27-restricted self-epitope correlates with ankylosing spondylitis. *J Clin Invest* 106: 47–53

29 McGonagle D, Gibbon W, Emery P (1998) Classification of inflammatory arthritis by enthesitis. *Lancet* 352: 1137–1140

30 Poole AR (1998) The histopathology of ankylosing spondylitis: are there unifying hypotheses? *Am J Med Sci* 316: 228–233

31 Maksymowych WP (2000) Ankylosing spondylitis at the interface of bone and cartilage. *J Rheumatol* 27: 2295–2301

32 Braun J, Khan MA, Sieper J (2000) Enthesitis and ankylosis in spondyloarthropathy: what is the target of the immune response? *Ann Rheum Dis* 59: 985–994

33 Zhang Y (2003) Animal models of inflammatory spinal and sacroiliac joint diseases. *Rheum Dis Clin North Am* 29: 631–645

34 Kuon W, Kuhne M, Busch DH, Atagunduz P, Seipel M, Wu P, Morawietz L, Fernahl G, Appel H, Weiss EH et al (2004) Identification of novel human aggrecan T cell epitopes in HLA-B27 transgenic mice associated with spondyloarthropathy. *J Immunol* 173: 4859–4866

35 Appel H, Kuon W, Kuhne M, Hulsmeyer M, Kollnberger S, Kuhlmann S, Weiss E, Zeitz

M, Wucherpfennig K, Bowness P et al (2004) The solvent-inaccessible Cys67 residue of HLA-B27 contributes to T cell recognition of HLA-B27/peptide complexes. *J Immunol* 173: 6564–6573

36 Zou J, Zhang Y, Thiel A, Rudwaleit M, Shi SL, Radbruch A, Poole R, Braun J, Sieper J (2003) Predominant cellular immune response to the cartilage autoantigenic G1 aggrecan in ankylosing spondylitis and rheumatoid arthritis. *Rheumatology (Oxford)* 42: 846–855

37 Zou J, Appel H, Rudwaleit M, Thiel A, Sieper J (2004) Analysis of the CD8+ T cell response to the G1 domain of aggrecan in ankylosing spondylitis. *Ann Rheum Dis* Nov 11; [Epub ahead of print]

38 Gao XM, Wordsworth P, McMichael A (1994) Collagen-specific cytotoxic T lymphocyte responses in patients with ankylosing spondylitis and reactive arthritis. *Eur J Immunol* 24: 1665–1670

39 Atagunduz P, Appel H, Kuon W, Wu P, Thiel A, Kloetzel PM, Sieper J (2005) HLA-B27-restricted CD8+ T cell response to cartilage-derived self peptides in ankylosing spondylitis. *Arthritis Rheum* 52: 892–901

40 May E, Dulphy N, Frauendorf E, Duchmann R, Bowness P, Lopez de Castro JA, Toubert A, Marker-Hermann E (2002) Conserved TCR beta chain usage in reactive arthritis; evidence for selection by a putative HLA-B27-associated autoantigen. *Tissue Antigens* 60: 299–308

41 Colbert RA (2004) The immunobiology of HLA-B27: variations on a theme. *Curr Mol Med* 4: 21–30

42 Allen RL, O'Callaghan CA, McMichael AJ, Bowness P (1999) Cutting edge: HLA-B27 can form a novel beta 2-microglobulin-free heavy chain homodimer structure. *J Immunol* 162: 5045–5048

43 Kollnberger S, Bird LA, Roddis M, Hacquard-Bouder C, Kubagawa H, Bodmer HC, Breban M, McMichael AJ, Bowness P (2004) HLA-B27 heavy chain homodimers are expressed in HLA-B27 transgenic rodent models of spondyloarthritis and are ligands for paired Ig-like receptors. *J Immunol* 173: 1699–1710

44 Boyle LH, Goodall JC, Opat SS, Gaston JS (2001) The recognition of HLA-B27 by human CD4(+) T lymphocytes. *J Immunol* 167: 2619–2624

45 Boyle LH, Hill Gaston JS (2003) Breaking the rules: the unconventional recognition of HLA-B27 by CD4+ T lymphocytes as an insight into the pathogenesis of the spondyloarthropathies. *Rheumatology (Oxford)* 42: 404–412

46 Yu D, Kuipers JG (2003) Role of bacteria and HLA-B27 in the pathogenesis of reactive arthritis. *Rheum Dis Clin North Am* 29: 21–36, v–vi

47 Penttinen MA, Heiskanen KM, Mohapatra R, DeLay ML, Colbert RA, Sistonen L, Granfors K (2004) Enhanced intracellular replication of Salmonella enteritidis in HLA-B27-expressing human monocytic cells: dependency on glutamic acid at position 45 in the B pocket of HLA-B27. *Arthritis Rheum* 50: 2255–2263

48 Sims AM, Wordsworth BP, Brown MA (2004) Genetic susceptibility to ankylosing spondylitis. *Curr Mol Med* 4: 13–20

49 Rudwaleit M, Höhler T (2001) Cytokine gene polymorphisms relevant for spondyloarthropathies. *Curr Opin Rheumtol* 13: 250–254

50 Laval SH, Timms A, Edwards S, Bradbury L, Brophy S, Milicic A, Rubin L, Siminovitch KA, Weeks DE, Calin A et al (2001) Whole-genome screening in ankylosing spondylitis: evidence of non-MHC genetic-susceptibility loci. *Am J Hum Genet* 68: 918–926

51 Zhang G, Luo J, Bruckel J, Weisman MA, Schumacher HR, Khan MA, Inman RD, Mahowald M, Maksymowych WP, Martin TM et al (2004) Genetic studies in familial ankylosing spondylitis susceptibility. *Arthritis Rheum* 50: 2246–2254

52 Martin TM, Zhang G, Luo J, Jin L, Doyle TM, Rajska BM, Coffman JE, Smith JR, Becker MD, Mackensen F et al (2005) A locus on chromosome 9p predisposes to a specific disease manifestation, acute anterior uveitis, in ankylosing spondylitis, a genetically complex, multisystem, inflammatory disease. *Arthritis Rheum* 52: 269–274

53 Hoyle E, Laval SH, Calin A, Wordsworth BP, Brown MA (2000) The X-chromosome and susceptibility to ankylosing spondylitis. *Arthritis Rheum* 43:1353–1355

54 Yin Z, Braun J, Neure L, Wu P, Liu L, Eggens U, Sieper J (1997) Crucial role of interleukin-10/interleukin-12 balance in the regulation of the type 2 T helper cytokine response in reactive arthritis. *Arthritis Rheum* 40: 1788–1797

55 Braun J, Yin Z, Spiller I, Siegert S, Rudwaleit M, Liu L, Radbruch A, Sieper J (1999) Low secretion of tumor necrosis factor alpha, but no other Th1 or Th2 cytokines, by peripheral blood mononuclear cells correlates with chronicity in reactive arthritis. *Arthritis Rheum* 42: 2039–2044

56 Rudwaleit M, Siegert S, Yin Z, Eick J, Thiel A, Radbruch A, Sieper J, Braun J (2001) Low T cell production of TNFalpha and IFNgamma in ankylosing spondylitis: its relation to HLA-B27 and influence of the TNF-308 gene polymorphism. *Ann Rheum Dis* 60: 36–42

57 Tuokko J, Koskinen S, Westman P, Yli-Kerttula U, Toivanen A, Ilonen J (1998) Tumour necrosis factor microsatellites in reactive arthritis. *Br J Rheumatol* 37:1203–1206

58 Hohler T, Schaper T, Schneider PM, Meyer zum Buschenfelde KH, Marker-Hermann E (1998) Association of different tumor necrosis factor alpha promoter allele frequencies with ankylosing spondylitis in HLA-B27 positive individuals. *Arthritis Rheum* 41: 1489–1492

59 Kaluza W, Leirisalo-Repo M, Marker-Hermann E, Westman P, Reuss E, Hug R, Mastrovic K, Stradmann-Bellinghausen B, Granfors K, Galle PR et al (2001) IL10.G microsatellites mark promoter haplotypes associated with protection against the development of reactive arthritis in Finnish patients. *Arthritis Rheum* 44: 1209–1214

60 Goedecke V, Crane AM, Jaakkola E, Kaluza W, Laiho K, Weeks DE, Wilson J, Kauppi M, Kaarela K, Tuomilehto J et al (2003) Interleukin 10 polymorphisms in ankylosing spondylitis. *Genes Immun* 4: 74–76

61 Howe HS, Cheung PL, Kong KO, Badsha H, Thong BY, Leong KP, Koh ET, Lian TY, Cheng YK et al (2005) Transforming growth factor beta-1 and gene polymorphisms in oriental ankylosing spondylitis. *Rheumatology (Oxford)* 44: 51–54

62 Jaakkola E, Crane AM, Laiho K, Herzberg I, Sims AM, Bradbury L, Calin A, Brophy S,

Kauppi M, Kaarela K et al (2004) The effect of transforming growth factor beta1 gene polymorphisms in ankylosing spondylitis. *Rheumatology (Oxford)* 43: 32–38

63   McGarry F, Neilly J, Anderson N, Sturrock R, Field M (2001) A polymorphism within the interleukin 1 receptor antagonist (IL-1Ra) gene is associated with ankylosing spondylitis. *Rheumatology (Oxford)* 40: 1359–1364

64   van der Paardt M, Crusius JB, Garcia-Gonzalez MA, Baudoin P, Kostense PJ, Alizadeh BZ, Dijkmans BA, Pena AS, van der Horst-Bruinsma IE (2002) Interleukin-1beta and interleukin-1 receptor antagonist gene polymorphisms in ankylosing spondylitis. *Rheumatology (Oxford)* 41: 1419–1423

65   Maksymowych WP, Reeve JP, Reveille JD, Akey JM, Buenviaje H, O'Brien L, Peloso PM, Thomson GT, Jin L, Russell AS (2003) High-throughput single-nucleotide polymorphism analysis of the IL1RN locus in patients with ankylosing spondylitis by matrix-assisted laser desorption ionization-time-of-flight mass spectrometry. *Arthritis Rheum* 48: 2011–2018

66   Timms AE, Crane AM, Sims AM, Cordell HJ, Bradbury LA, Abbott A, Coyne MR, Beynon O, Herzberg I, Duff GW et al (2004) The interleukin 1 gene cluster contains a major susceptibility locus for ankylosing spondylitis. *Am J Hum Genet* 75: 587–595

67   Djouadi K, Nedelec B, Tamouza R, Genin E, Ramasawmy R, Charron D, Delpech M, Laoussadi S; EUROAS (2001) Interleukin 1 gene cluster polymorphisms in multiplex families with spondylarthropathies. *Cytokine* 13: 98–103

68   Jin L, Zhang G, Akey JM, Luo J, Lee J, Weisman MH, Bruckel J, Inman RD, Stone MA, Khan MA et al (2004) Lack of linkage of IL1RN genotypes with ankylosing spondylitis susceptibility. *Arthritis Rheum* 50: 3047–3048

69   Crane AM, Bradbury L, van Heel DA, McGovern DP, Brophy S, Rubin L, Siminovitch KA, Wordsworth BP, Calin A, Brown MA (2002) Role of NOD2 variants in spondylarthritis. *Arthritis Rheum* 46: 1629–1633

70   Peeters H, Vander Cruyssen B, Laukens D, Coucke P, Marichal D, Van Den Berghe M, Cuvelier C, Remaut E, Mielants H et al (2004) Radiological sacroiliitis, a hallmark of spondylitis, is linked with CARD15 gene polymorphisms in patients with Crohn's disease. *Ann Rheum Dis* 63: 1131–1134

71   van der Paardt M, Crusius JB, de Koning MH, Murillo LS, van de Stadt RJ, Dijkmans BA, Pena AS, van der Horst-Bruinsma IE (2003) CARD15 gene mutations are not associated with ankylosing spondylitis. *Genes Immun* 4: 77–78

72   Ferreiros-Vidal I, Amarelo J, Barros F, Carracedo A, Gomez-Reino JJ, Gonzalez A (2003) Lack of association of ankylosing spondylitis with the most common NOD2 susceptibility alleles to Crohn's disease. *J Rheumatol* 30: 102–104

73   Brown MA, Edwards S, Hoyle E, Campbell S, Laval S, Daly AK, Pile KD, Calin A, Ebringer A, Weeks DE et al (2000) Polymorphisms of the CYP2D6 gene increase susceptibility to ankylosing spondylitis. *Hum Mol Genet* 9: 1563–1566

74   Yen JH, Tsai WC, Chen CJ, Lin CH, Ou TT, Hu CJ, Liu HW (2003) Cytochrome P450 1A1 and manganese superoxide dismutase genes polymorphisms in ankylosing spondylitis. *Immunol Lett* 88: 113–116

75   van der Paardt M, Crusius JB, de Koning MH, Morre SA, van de Stadt RJ, Dijkmans BA, Pena AS, van der Horst-Bruinsma IE (2005) No evidence for involvement of the Toll-like receptor 4 (TLR4) A896G and CD14-C260T polymorphisms in susceptibility to ankylosing spondylitis. *Ann Rheum Dis* 64: 235–238

76   Jin L, Weisman M, Zhang G, Ward M, Luo J, Bruckel J, Inman R, Khan MA, Schumacher HR, Maksymowych WP et al (2005) Lack of association of matrix metalloproteinase 3 (MMP3) genotypes with ankylosing spondylitis susceptibility and severity. *Rheumatology (Oxford)* 44: 55–60

77   Timms AE, Zhang Y, Bradbury L, Wordsworth BP, Brown MA (2003) Investigation of the role of ANKH in ankylosing spondylitis. *Arthritis Rheum* 48: 2898–2902

78   Tsui HW, Inman RD, Paterson AD, Reveille JD, Tsui FW (2005) ANKH variants associate with ankylosing spondylitis: gender differences. *Arthritis Res Ther* 7: R513–R525

D. Tools for analysing complexity

Emerging tools for dissecting complex disease

Ulf D. Landegren

Department of Genetics and Pathology, The Rudbeck Laboratory, Uppsala University, Dag Hammarskjoldsvag 20, 75185 Uppsala, Sweden

Introduction

Like the proverbial elephant being felt by a group of blind men, human diseases are perceived differently depending on how investigating scientists are able to study them. Through the years notions of alterations of body humors; autotoxicity/auto-immunity; influences by the environment, upbringing or diet; psychosomatics; infectious agents; or genetic factors have all taken turn dominating thinking about the origin of diseases, depending on available research technologies. In recent years the seductive example of genetic linkage analysis of monogenic disease, combined with the recent availability of total genome sequence information, have motivated the search for genetic factors also in complex diseases where many genes may contribute to predisposition, along with a range of nonheritable factors. The availability of comprehensive lists of genes and of genetic variants, as well as of proteins and their variants, along with high-throughput approaches to study them in extensive collections of biobanked patient samples and in model organisms, should over coming years yield profound insights in disease aetiologies whether genetic, infectious, or otherwise. This can be justifiably expected to enable earlier, even presymptomatic diagnosis, molecularly directed therapies, and improved follow up. Also without a real understanding of molecular mechanisms underlying disease, the ability to identify diagnostic molecular patterns before extensive organ damage occurs, or to distinguish phenocopies of current disease entities that may require distinct therapies, promises to improve treatment or maybe allow diseases to be avoided outright. In this chapter I will briefly discuss some modalities of molecular analysis, before describing tools being developed in our lab for molecular analysis in research, diagnostics, and for monitoring of disease.

Accessing increasing amounts of molecular information

Greatly improved technologies are becoming available to follow the flow of biological information from its repository in our genomes, over sequences expressed as

The Hereditary Basis of Rheumatic Diseases, edited by Rikard Holmdahl
© 2006 Birkhäuser Verlag Basel/Switzerland

RNA, and into protein, and then functional effects of these [1]. Some technologies are geared to provide an overview of ongoing processes like gene expression at levels of RNA and protein. Since only a finite number of cellular states are possible, scoring of a limited number of gene products should suffice to distinguish different functional conditions of tissues in order to identify biomarker patterns that are diagnostic for specific diseases, or for responses to therapy. By contrast, for mechanistic understanding of molecular processes such as in the course of systems biological analyses where an ultimate goal is to be able to compute biological processes, the requirements for comprehensiveness and precision are significantly greater.

The founder of the microprocessor company Intel, Gordon Moore, in 1965 formulated what became known as Moore's law stating that the density of transistors on microchips, and thus their capacity for storing and processing information, would continue to approximately double every year and a half. The ability to acquire and process molecular genetic information continually undergoes a development analogous to that of our computers – and for related reasons.

DNA sequencing

Since the 1970s, the rate of DNA sequence acquisition has closely followed the increase of computational information processing, and it proved possible to sequence the entire human genome, encompassing more than 3 billion nucleotides by the beginning of the new millennium. After this further progress has been made and a commonly stated goal over the next few years is to now develop methods that will allow an individual's total genome, or the altered version thereof in a tumour let's say, to be sequenced for a mere 1,000 dollars/euros. Several promising approaches, distinct from the classical Sanger sequencing method, are currently in development [2]. Once perfected, they may take the place of many current methods for DNA analysis.

Genotyping

Awaiting such breakthroughs, maps have been prepared of all those places where human genomes commonly differ from one individual to another, in order to focus molecular analyses on particularly important features. Methods to analyse such variants, generally referred to as genotyping, have improved greatly. Only a few years back investigations were done painstakingly one locus and one DNA sample at a time, and this is indeed how analyses are still done in most labs. However, now there is an assortment of methods that allow tens to hundreds of thousand genetic variants to be assessed in parallel in individual DNA samples in the search for disease-associated genetic variants, which is genetic variants whose frequencies appear to differ significantly between groups of patients and closely matched control subjects. Of key importance for this development has been the elaboration of methods to interrogate large numbers of specific nucleotide sequences. One such approach for high-throughput genetic analyses will be described later in this chapter.

Gene copy number measurements

Methods similar to those used for association studies can also be applied to investigate copy numbers of sequences across the genome, particularly in children with congenital defects that may be caused by chromosomal imbalances, and the genomic resolution of such analyses increases rapidly. Differences between genomes with respect to copy numbers of specific sequences have emerged as a considerable source of genetic variation. Copy number differences of genomic regions are also frequently observed in tumour tissues compared to normal tissues from the same individual, offering insights into what genes may have been amplified in copy numbers or lost during the progress of the tumours.

Transcript copy number measurements

Transcript copy number measurements provide a perspective on which gene programs are currently running in an investigated tissue, thus reflecting the state of differentiation and activity of the cells. Also this mode of analysis continues to evolve towards greater sensitivity and precision, and by some degree of standardisation that permits comparison between investigations in different labs and using different methods. The methods allow for testing increasing numbers of features of transcriptomes, such as splice variation or the expression of genes without apparent protein coding potential. There is a desire to push detection sensitivities to permit RNA expression analyses of single cells in order not to have to average analyses over many cell types and cell cycle phases in a tissue sample, as this will obscure gene programs in individual cells.

It is also of interest, but so far not done on a large scale, to investigate differential expression of allelic variants of genes. Besides revealing differences of expression of paternally and maternally inherited genes subject to genomic imprinting, such analyses can also serve to demonstrate functional differences among allelic promoter variants, and loss or amplification of genomic regions in tumour tissue. Furthermore, allele-specific expression can pin-point transcripts that harbour nonsense mutations in, e.g., tumour tissue, exploiting the circumstance that transcripts with illegitimate stop codons are subject to a mechanism referred to as nonsense-mediated decay, leading to reduced copy numbers because of premature degradation of transcripts of such alleles.

A possible alternative means of assessing global patterns of gene regulation is by studying the pattern of methylation of sequence motifs involving the dinucleotide CG in genomic DNA, since this tends to reflect transcriptional activity of nearby genes in the investigated tissue. Such analyses may present some advantages over RNA measurements as they can be performed using the increasingly efficient genotyping methods, they concern DNA rather than unstable RNA molecules, and levels of methylation varies between 0 and 100%, allowing for simplified quantitation compared to RNA expression levels that can vary over four or more orders of magnitude.

Protein measurements

For a number of reasons, investigations of protein expression are potentially even more revealing than assessment of levels of transcripts of different genes: Proteins are probably more directly involved in cellular functions, and their expression is subject to regulation not reflected at the level of the corresponding mRNA molecules. Moreover, effects of processing, covalent modification by phosphorylation or other mechanisms, redistribution within cells, and the dynamics of interactions among proteins can only be assessed at the protein level. It is also of importance that unlike RNA, which must be studied in the tissues where the genes are expressed, sufficiently sensitive protein detection methods could allow analyses via blood or other body fluids also when the affected organ is inaccessible or indeed unknown. This carries with it the potential that it may be possible to diagnose diseases at early, still treatable stages by simple blood tests that can be repeated on a yearly basis. Recent developments of mass spectrometry-based techniques have greatly improved opportunities for extensive protein analyses, but as illustrated further down we can also look forward to increasingly specific, sensitive, and comprehensive methods for protein analyses exploiting specific affinity reagents like antibodies and related reagents.

Other molecular analyses

Numerous molecular analyses are currently being scaled-up for extensive characterisation of biological systems in basic research. Some of these may assume important roles also in the search for mechanisms underlying disease, and to identify diagnostic or therapeutic targets. For example, by isolating genomic DNA fragments bound by specific transcription factors, and then identifying these sequences by hybridisation to DNA microarrays using the so-called ChIP on Chip-technique, sequences can be identified that are regulated by specific DNA factors in a given tissue. Also this analysis could provide valuable perspectives on disease processes by revealing the transcriptional state of affected tissues. There is a growing interest also to map interactions between different proteins, and to study these interactions as a function of the state of the investigated tissues, since the dynamics of protein interactions may well prove as informative, perhaps even more so than the expression levels of individual proteins. Also co-localisation between specific genomic regions may provide information on the functional states of tissues.

In situ analyses

One further analytical modality that is likely to gain increasing importance in research as well as in diagnostics is the imaging of specific molecules directly in tissues and cells, ultimately allowing routine analyses of any individual molecules and of large sets of molecules. Such analyses promise to contribute greatly to improved mechanistic understanding of cellular functions, and they can reveal aspects of disease processes that fail to be represented as alterations of bulk numbers of molecules, averaged over many cells in a tissue sample. Molecular imaging will also assist in routine pathological examination as well as in high-content screening for effects by drug candidates in the course of drug development. Significantly improved technological opportunities can be anticipated in all these modes of analyses, allowing very large numbers of target molecules to be assessed with single-molecule sensitivity. In the following I will briefly describe a family of procedures for molecular investigations, currently being developed in our lab.

A set of DNA ligation-based techniques for molecular analyses

An overarching aim in our lab is to develop tools that can enable the comprehensive molecular analyses described in this chapter of all classes of macromolecules. The tools must exhibit excellent specificity and sensitivity, and be possible to implement for large numbers of molecules in parallel, allowing even single target molecules to be identified directly in biological samples. For this purpose we have used as a general strategy the joining by enzymatic ligation of strands of probe DNA molecules. DNA ligation allows coincident detection by pairs of probes to yield amplifiable joined DNA strands that can include signature DNA sequences, serving to represent the detected target molecule. Some valuable principal features of ligase-based analyses are listed in Table 1.

The oligonucleotide ligation assay

An early method to detect, distinguish, and to quantify target nucleic acid sequences – the oligonucleotide ligation assay (OLA) – involves the joining of pairs of oligonu-

Table 1 - Some principal advantages of ligase-based analyses

Pairwise probe hybridisation to target molecules > High specificity of detection
Enzyme-based substrate recognition > High selectivity of detection
Affinity interaction matures into covalent interaction > Efficient linking of probes
Formation of new DNA strand > Amplifiable reaction product
Inclusion of sequence motifs > Product carries information for identification purposes

Figure 1
The oligonucleotide ligation assay
Two oligonucleotides are designed to hybridise to adjacent sequences on a target DNA strand. If the juxtaposed ends are properly hybridised the two oligonucleotides can be joined by ligation generating a detectable molecular species. Even a single mismatch at one of the ends to be joined reduces ligation efficiency by at least hundred-fold.

cleotide probes hybridising in juxtaposition on a target DNA strand (Fig. 1) [3]. Mismatched probes fail to be joined by the ligase enzyme due to its substrate requirements, allowing sequence variants to be conveniently distinguished. The circumstance that two adjacent target sequences must be recognised ensures very high detection specificity, sufficient to specifically detect unique target sequences directly in the human genome.

Padlock probes

A modified version of OLA involves the use of oligonucleotides, the two ends of which can be joined by ligation upon hybridising to adjacent target sequences in a DNA sample, so-called padlock probes (Fig. 2) [4]. These probes share many features with the OLA procedure, but they also exhibit additional valuable properties: The fact that the two target-complementary segments are part of the same

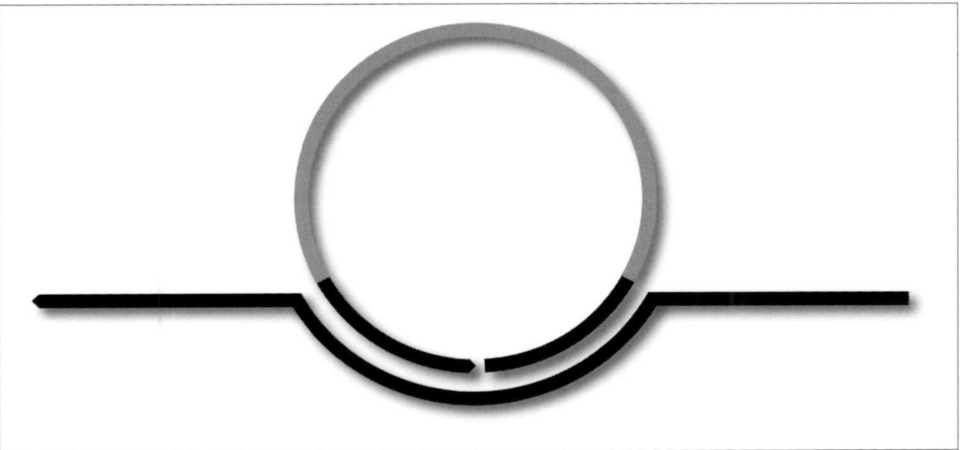

Figure 2
Padlock probe ligation
Linear oligonucleotides are designed so that their 5' and 3' ends can hybridise to adjacent sequences on a target strand. Upon addition of a ligase the two ends are covalently joined, resulting in a circular DNA strand.

probe molecule allows successful reaction products to be identified as circular DNA strands, easily distinguished from remaining unreacted probes or spurious ligation products arising through joining of pairs of probes. This property also allows very large numbers of probes to be combined without a concomitantly increased risk of incorrect detection signals from crossreactive products [5], something that has limited the scaling-up of PCR for analysis of multiple target sequences.

The circumstance that reacted probes form DNA circles also has valuable consequences for detection. Due to the double-helical nature of the hybridised and ligated probes the DNA circles become linked – catenated – to their target sequences. This means that washing stringency need not be limited by the hybridisation stability of the probes, as the probes are in fact bound to their target molecules like links in a chain.

Rolling-circle replication of reacted probes

Of greater importance still; circularised probes can template a very useful amplification reaction by being copied in a so-called rolling circle replication process, where a DNA polymerase replicates the endless DNA template like a cat chasing its

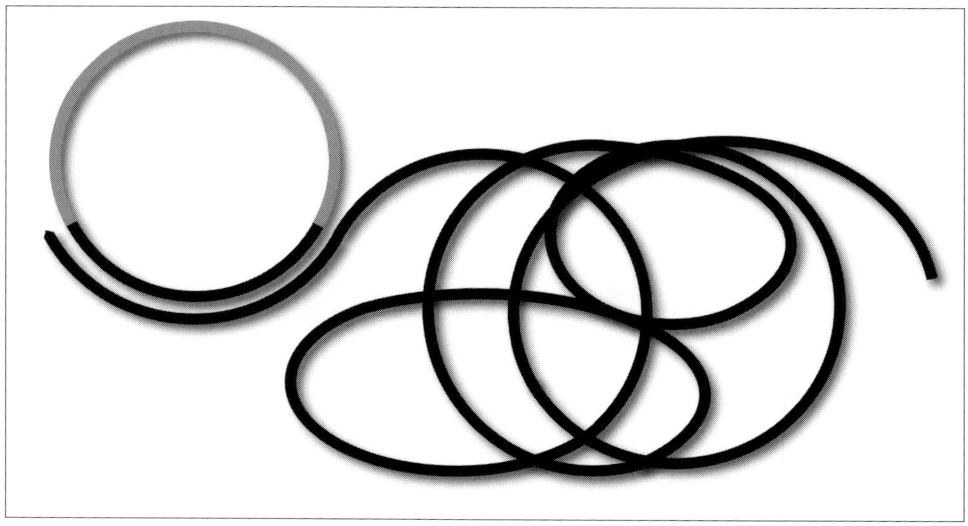

Figure 3
Rolling-circle replication
A circular DNA strand, which can have been formed by ligation of a padlock probe or through proximity ligation, can be replicated in a rolling-circle replication process, resulting in a long, single stranded product composed of tandem repeats of the complement of the circularised strand. Such amplified molecules are easily visualised using a detectable probe that hybridises to each repeat of the amplification product.

tail (Fig. 3) [6]. Replication products form long, contiguous DNA strands composed of many copies of the probes, including any signature sequences incorporated in the probes for identification purposes. Such covalently linked amplification products are highly unlikely to arise in the absence of proper probe circularisation, and even single concatamers composed of, say, 1,000 probe copies, give rise to easily detectable signals by hybridisation with fluorescent probes. The combination of padlock circularisation and probe replication therefore permits *in situ* detection of even single target DNA molecules, something that is of great potential value in analysing tissue heterogeneity [7].

Proximity ligation

Recently, we have demonstrated that the probe ligation mechanism need not necessarily be limited to the detection of nucleic acid sequences. In proximity ligation assays a short DNA strand is added that can template the enzymatic ligation of free

Figure 4
Proximity ligation
Two affinity reagents, for example antibodies, binding the same target molecule, are both equipped with short oligonucleotides. Upon coincident binding to the target molecule, the ends of the two oligonucleotides are brought in proximity and can be joined by ligation, templated by an added connector oligonucleotide, giving rise to an amplifiable DNA strand.

ends of oligonucleotides attached to specific affinity reagents (Fig. 4) [8–10]. Thereby, amplifiable information carrying DNA strands can be generated that serve as surrogate markers for any molecular targets bound by two of the affinity reagents. The specificity of such reactions depends on the choice of affinity probes. If pairs of antibodies are used that are directed at different determinants on the same target proteins, or on interacting proteins, then coincident binding of pairs of antibodies with attached DNA strands to individual or pairs of proteins can give rise via DNA ligation to amplifiable DNA strands. Through suitable assay design very low non-specific signals can be achieved, while specific detection reaction products can be amplified by PCR with real-time detection, or through rolling-circle replication, both resulting in extremely sensitive protein detection.

The assays can be performed either as convenient homogenous detection reactions with no need for washes, or in a solid-phase format, allowing interfering sample components and excess reagents to be removed by washes before the ligation reaction. It is also possible to configure the assays so that formation of an amplifiable product requires binding by sets of three or more affinity probes for further enhanced specificity of recognition, and for further background reduction.

Since it is possible to control which pairs of detection reagents that can be joined by ligation, the assay should be suitable for multiplex protein detection, avoiding any risks of cross reactivity. We have also demonstrated that the proximity ligation mechanism for protein detection, just like padlock probes for nucleic acid detection, can be combined with rolling circle replication for *in situ* detection of even single protein molecules, or interacting pairs of molecules.

Towards ligation-based high-throughput biology

As discussed in this chapter, a very large number of DNA and RNA sequences are being identified as targets of interest for analysis using specific DNA probes such as padlock probes. It is now important to develop probe production techniques that allow for parallel synthesis of thousands of probes at costs significantly below present levels. For protein detection there is rapidly increasing interest in different forms of analyses, including detection of exceedingly low concentrations of specific proteins disseminated in the blood stream, and proximity ligation may provide the required sensitivity and properties for parallel analyses. Protein detection differs from nucleic acid analyses, however, because of the added difficulty of generating specific affinity reagents, since probe design cannot be based on simple complementarity rules. This obstacle is rapidly being removed because of ongoing efforts to raise complete sets of affinity probes, antibody-based or otherwise, against all human proteins and their variants.

The ligase-based tool kit described herein thus has the potential to serve as a general solution for highly multiplexed analyses of any and all macromolecules for both quantitation and localisation purposes, and the specificity of the reactions allow even single target molecules to be identified directly in cells and tissues. Such a uniform set of probing mechanisms and assay formats could contribute to harmonise global efforts to chart the molecular basis of complex disease, and also support the application of this knowledge in future diagnostics approaches.

Acknowledgements
The work in our laboratory is supported by grants from the EU (the FP6 Integrated Project 'MolTools'), by the Wallenberg Foundation, and from the Swedish Research Councils for natural sciences and for medicine.

References

1   Sauer S, Lange BM, Gobom J, Nyarsik L, Seitz H, Lehrach H (2005) Miniaturization in functional genomics and proteomics. *Nat Rev Genet* 6: 465–476

2   Shendure J, Mitra RD, Varma C, Church GM (2004) Advanced sequencing technologies: methods and goals. *Nat Rev Genet* 5: 335–344

3   Landegren U, Kaiser R, Sanders J, Hood L (1988) A ligase-mediated gene detection technique. *Science* 241: 1077–1080

4   Nilsson M, Malmgren H, Samiotaki M, Kwiatkowski M, Chowdhary BP, Landegren U (1994) Padlock probes: circularizing oligonucleotides for localized DNA detection. *Science* 265: 2085–2088

5   Hardenbol P, Baner J, Jain M, Nilsson M, Namsaraev EA, Karlin-Neumann GA, Fakhrai-Rad H, Ronaghi M, Willis TD, Landegren U, Davis RW (2003) Multiplexed genotyping with sequence-tagged molecular inversion probes. *Nat Biotechnol* 21: 673–678

6   Baner J, Nilsson M, Mendel-Hartvig M, Landegren U (1998) Signal amplification of padlock probes by rolling circle replication. *Nucleic Acids Res* 26: 5073–5078

7   Larsson C, Koch J, Nygren A, Janssen G, Raap AK, Landegren U, Janssen G, Raap AK, Landegren U, Nilsson M (2004) *In situ* genotyping individual DNA molecules by target-primed rolling-circle amplification of padlock probes. *Nat Methods* 1: 227–232

8   Fredriksson S, Gullberg M, Jarvius J, Olsson C, Pietras K, Gustafsdottir SM, Olsson C, Pietras K, Gustafsdottir SM, Ostman A, Landegren U (2002) Protein detection using proximity-dependent DNA ligation assays. *Nat Biotechnol* 20: 473–477

9   Gullberg M, Fredriksson S, Taussig M, Jarvius J, Gustafsdottir S, Landegren U (2003) A sense of closeness: protein detection by proximity ligation. *Curr Opin Biotechnol* 14: 82–86

10 Gullberg M, Gustafsdottir SM, Schallmeiner E, Jarvius J, Bjarnegard M, Betsholtz C, Landegren U, Fredriksson S (2004) Cytokine detection by antibody-based proximity ligation. *Proc Natl Acad Sci USA* 101: 8420–8424

Expression analysis of rheumatic diseases, prospects and problems

Thomas Häupl[1], Andreas Grützkau[2], Joachim Grün[3], Andreas Radbruch[2] and Gerd Burmester[1]

[1]Department of Rheumatology, Charité, University Hospital, Tucholskystraße 2, 10117 Berlin, Germany; [2]German Arthritis Research Centre, Berlin, Germany; [3]Oligene GmbH, Berlin, Germany

Introduction

Inflammatory rheumatic diseases are among the greatest diagnostic challenges in modern medicine. Pathognomonic markers for obvious identification as well as direct markers for disease activity and therapeutic responsiveness in the chronic stage are missing because most of these diseases are etiopathologically undefined and patient assessment relies on autoimmune phenomena, organ diagnostics and unspecific parameters of systemic inflammation. In light of these limitations in clinical rheumatology, expectations in genomics are high.

Technically, gene expression profiling has opened new avenues. Instead of single or a handful of candidates, tens of thousands of different genes can be investigated at one time. Transcriptome analysis is currently the most advanced and comprehensive approach to screen genome-wide for gene activity. Although at a slower pace, proteome analyses are also rapidly improving and will provide a deeper insight beyond the capabilities of transcriptome information. Furthermore, genome mutations predisposing for rheumatic diseases may help in primary diagnosis as well as prognosis of the disease [1].

The aims of molecular analyses in rheumatic diseases are clearly defined and focus on 1) markers for early diagnosis, 2) parameters to define severity and activity of the disease which may identify subtypes or leading pathomechanisms, 3) parameters for prognosis, and 4) markers for therapeutic stratification and candidates for therapeutic intervention. In principle, this may be possible only with diagnostic pattern devoid of a functional context. However, only understanding the cellular and molecular mechanisms of chronic inflammation has improved the therapeutic armamentarium with biologics. Therefore, our ultimate goal has to be the functional interpretation of expression results. To achieve this goal, systematic analyses, collating of information and development of molecular network models will be essential.

The Hereditary Basis of Rheumatic Diseases, edited by Rikard Holmdahl
© 2006 Birkhäuser Verlag Basel/Switzerland

Expression analysis in rheumatic diseases

The first report on array hybridisation in rheumatic diseases was published by Heller et al. [2]. A customised array of 96 immunologically relevant genes was applied to demonstrate that multiparameter screens may be useful in the analysis of inflammatory diseases such as rheumatoid arthritis. Several matrix metalloproteines (MMPs), MMP inhibitors, cytokines and chemokines were detectable in rheumatoid synovium.

In subsequent studies with increasing numbers of genes beyond 10,000 and technical improvement in terms of reliability, arrays were more and more applied for gene discovery and diagnostic pattern identification. Maas et al. [3] suggested an algorithm for discriminating autoimmune from normal immune responses by profiling peripheral blood mononuclear cells (PMBC) from patients with rheumatoid arthritis, systemic lupus erythematosus (SLE), type I diabetes and multiple sclerosis (MS) compared to PBMC from healthy donors after influenza vaccination. Their candidates were involved in apoptosis, cell cycle progression, cell differentiation and cell migration. Relatives of the patients with autoimmune diseases were also sorted to the disease group. Thus, the authors speculate that their gene selection may reflect a genetic trait rather than the different disease processes, which could not be distinguished from each other.

Studies on rheumatoid synovium by van der Pouw Kraan and co-workers [4, 5] which will be further discussed in another chapter, were aiming for functionally relevant genes and for the identification of RA subgroups to improve diagnostic and therapeutic stratification. Devauchelle and co-workers [6] demonstrated that a set of 48 genes was able to classify rheumatoid arthritis (RA) from non-RA. Aiming for discovery of new candidates, they selected six genes for further confirmation by polymerase chain reaction (PCR) in the tissue and three in synovial fibroblast cultures.

Also tissue-based, the approach by Morawietz et al. [7] showed various inflammatory genes to be upregulated in chronic inflammation of periprosthetic membranes of RA and osteoarthritis (OA) patients in the process of prosthetic loosening.

PBMC were again the target of investigation in studies by Gu et al. [8] in spondyloarthropathies, rheumatoid arthritis, and psoriatic arthritis. Their dominant candidates included several genes which were inflammation related including CXCR4. As SDF-1, the ligand of CXCR4, was found increased in the synovial fluids of arthritides, an important role of this chemotactic axis in SpA and RA was suggested.

Interesting studies on functional signatures were performed by several groups in systemic lupus on PBMC [9–12]. One group [10] confirmed these findings by the experimental proof that 23 of their 161 candidates were induced by IFN-α, -β or -γ in PBMC from healthy donors. Bennett et al. [9] also identified genes involved in granulopoiesis, which belonged to cells of the myeloic lineage that were co-separated only in SLE.

In an initial attempt to address the problems of mixed cell populations and unspecific dilution of low expressed candidates in tissues as well as to allow histological association of complete profiles, Judex et al. [13] demonstrated the feasibility of microarray analysis from laser microdissected areas of synovial tissues with a minimum of 600 cells.

Pierer et al. [14] circumvented the problem by performing functional analyses on purified synovial fibroblasts *in vitro*. They stimulated via toll-like receptor 2 and investigated the chemokine response. On the other hand, Schmutz et al. [15] directly focused on the investigation of chemokine receptor expression by profiling whole tissue and suggested CXCR5 as relevant in RA.

The studies summarised above were selected as a representation of typical array experiments performed from patient material either *ex vivo* as unsorted complex tissues, as blood cells composed of different populations or as *in vitro* stimulation experiment. All these studies have identified genes, which could be possible candidates for further analysis, diagnostic application or therapeutic targeting.

Problems

Statistical analysis

The major problem for all research groups is the management of the flood of data and the interpretation of data in a reasonable and unbiased way. Usually, a very small number of samples are analysed compared to the extremely large number of genes. For such small sample sizes, there is a 100% chance to select candidates only by chance. Furthermore, gene selection depends on the underlying strategy, hypothesis or discovery driven.

Hypothesis driven functional studies are often aiming for precise groups of candidate genes and thereby restrict data analysis to genes of interest. Discovery driven comparative studies with complex clinical samples are extremely difficult to analyse because optimised bioinformatic tools are insufficiently developed. For example, in the field of cancer, the seven major studies published on prediction of outcome were reanalysed with more advanced algorithms [16]. Five of these seven studies did not classify patients better than chance. We may therefore be cautious with results that were extracted from microarray data in the field of rheumatology as these new algorithms were not applied in most of these studies.

Furthermore, comparison between different studies is difficult. Different platforms cannot be readily compared and data comparison is often restricted to the number of genes that were published. Therefore, current analyses give first suggestions for candidate genes. However, it is impossible to interpret comprehensively the complexity of the transcriptome in the biological systems that were investigated, although, genome-wide microarrays may be the tools eventually providing such information.

Mixed cellular composition

A major challenge with respect to comprehensive interpretation is the variable composition of different cell types in a given sample of inflamed tissue or peripheral blood. All discovery driven studies referenced above were performed with such mixed samples. It is a problem of almost every clinical sample in any type of disease, including tumour tissues [17–20].

We have compared the impact of disease on one specific cell type with the differences between a normal cell type and a normal tissue. Figure 1 demonstrates that there are many more genes and to a higher extent differentially expressed, for example between normal monocytes and normal synovium, than between normal monocytes and rheumatoid arthritis monocytes.

Variability in cellular composition will therefore have a major impact on the quality and quantity of differentially expressed genes. This will increase statistical variability, will be more prone to false selection in cases of a small number of samples and will therefore demand for larger sample sizes as suggested for the cancer studies [16]. Another consequence of mixed cellular composition is that genes related to infiltration of a cell type may be preferably selected, while genes involved in the molecular processes of the disease may be lost in the majority of minor differences. As a third aspect, we have to consider that many genes provide their function not by upregulation of transcription but by protein modification. Thus, the message will be detectable in a normal and un-stimulated cell. If such a gene is known as a potent signal transmitter, infiltration of the corresponding cell type will pretend active participation of the gene in the disease process, although, this may not be true. Finally, mixed cell types dilute the message of a single cell type and rare transcripts may fall below the threshold of detection.

These problems are inherent to all studies performed with samples of complex cellular composition. In the different studies referenced above, only one single group [9] tried to address this problem and identified differential cellular composition as the source of a certain panel of differentially expressed genes. When we compared the list of candidates published, for example in another study on PBMC [8], with our own expression profiles of different highly purified subtypes of peripheral blood cells, all genes were found to be expressed highest in neutrophils. In light of the first study [9] that such cells may be co-separated in inflammatory diseases, the data published by the second study [8] have to be considered with caution and need further confirmation.

Differential expression and biological importance

Statistical measures determine the variability between samples. Intensity of signals and signal differences influence this measure. Signals reflect the abundance of

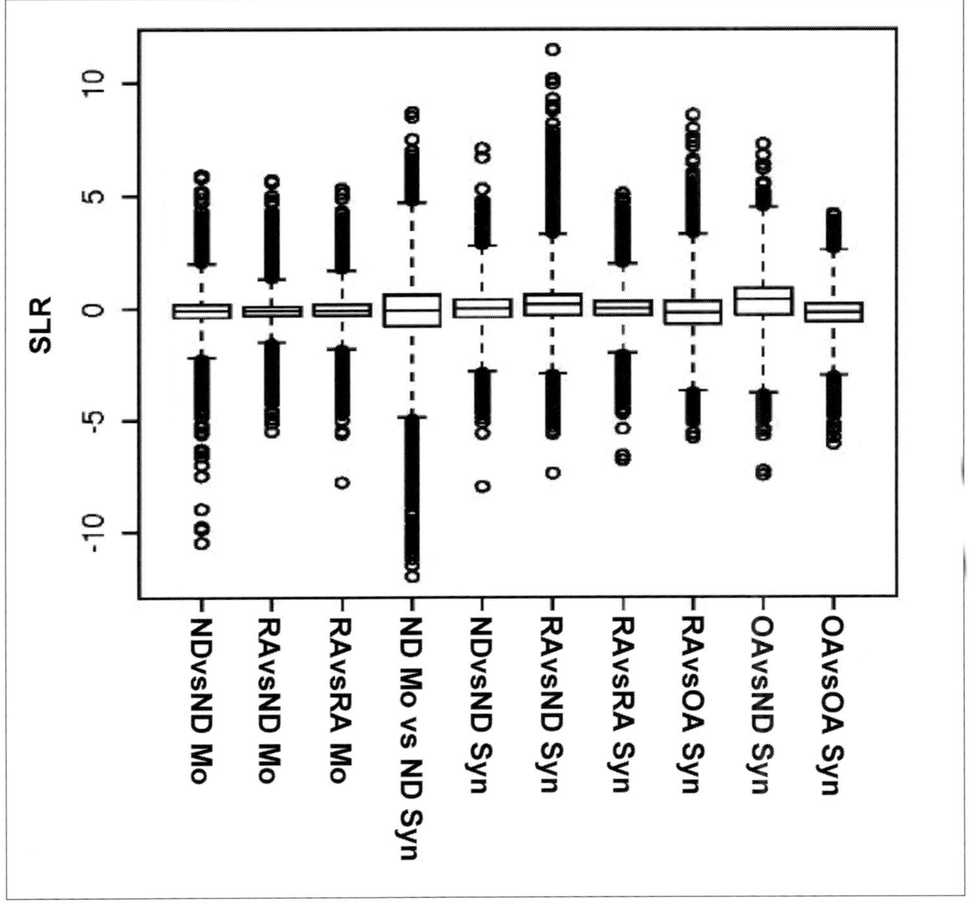

Figure 1
Signal log ratios (SLR) of expression comparison analysis between different normal donors (ND) or patients with rheumatoid arthritis (RA) using monocytes from peripheral blood (Mo) or synovial tissues (Syn). Expression was compared with MAS 5.0 software (Affymetrix). The distribution of SLR values by box-plots demonstrates that comparison of different cell types from the same donor group (ND Mo vs ND Syn) reveals more and higher differences than comparison between the same cell type (Mo or Syn) from different individuals from the same (ND vs ND or RA vs RA) or even different donor groups (RA vs ND).

mRNA in a given sample. This abundance is related to the functional category to which a gene belongs. And functional importance does not always relate to the intensity of signal or signal differences. Figure 2 demonstrates such differences in

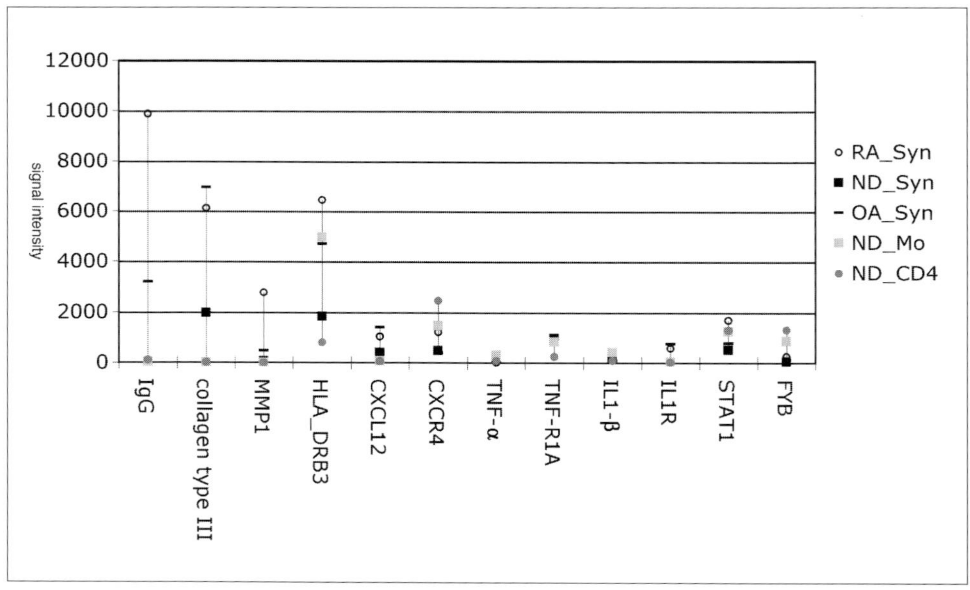

Figure 2

Expression signals of different genes in different cell types. Different molecular functions (extracellular or intracellular action, martrix production or signalling) require different absolute quantities, or changes of quantities upon activation. IgG, immunoglobulin G; MMP1, matrix metalloproteinase 1; HLA, human leukocyte antigen; CXCL12, chemokine SDF1; CXCR4, receptor for SDF1; TNF-α, tumor necrosis factor alpha; IL1-β, interleukin-1 beta; IL1R, receptor for IL1-β; STAT1, signal transducer and activator of transcription 1; FYB, FYN binding protein; RA, rheumatoid arthritis; ND, normal donor; OA, osteoarthritis; Mo, monocytes; CD4, CD4 positive T-cells; Syn, synovial tissue.

signal intensities related to function for representative genes of different categories, extracellular effectors and extra- and intracellular signalling.

Extracellular effector molecules like immunoglobulins and matrix molecules (collagens) are needed in high abundance and are therefore represented in the array with high signal intensities. Proteases involved in matrix turnover can cleave more than one matrix molecule. This may explain that expression levels are very low in normal tissue and are significantly increased only in exacerbating destructive processes like in RA.

Signalling molecules are expressed at different levels depending on 1) extracellular or intracellular location, 2) long or short distance action, 3) pre-formation and activation on the protein level or newly transcribed and synthesised as immediately active molecules. Chemokines like SDF1 are involved in extracellular long distance signalling and may become newly transcribed and highly expressed when needed. In contrast, the corresponding receptor CXCR4 is pre-formed on the surface of normal cells. High numbers of receptor molecules may compensate for smaller concentrations of signalling molecules which are diluted with increasing distance from the producer cell. Intensities measured for the receptor may even exceed those for the signalling molecule in a mixed sample. Short acting proinflammatory cytokines like TNF alpha and IL-1, which signal for danger and act in small concentrations, are strictly controlled. This may explain the low intensities and the difficulties to detect such important players in the inflammatory process. This problem may be aggravated by the analysis of mixtures of different cells, which may further dilute the concentration of low abundant messages. In contrast, intracellular signalling molecules like STAT1 or FYB are preformed molecules regulated on the protein level and may become transcribed higher only to a small extent as intracellular concentration will quickly increase and multiply intracellular effects.

These dependencies between biological importance and mathematical selection process are very critical and need to be considered for interpretation of expression results. They may explain why TNF inhibition is a successful therapy in arthritis although array results would not point to this cytokine as the key player [5]. Similarly, no interferon was identified in the SLE studies. In fact, the activity of such signalling molecules may be much easier identified by determining the signatures which are induced by these molecules [10].

Prospects

Statistical analysis of microarrays has significantly improved within recent years. The homepage www.bioconductor.org provides a broad panel of highly sophisticated tools for research. These are selecting for the best and most stable panel of genes from a given analysis by different statistical and iteratively testing algorithms. Compared to the problems of the mixed cellular composition in clinical samples, the divergence of numerical and biological importance, and the lack of molecular network knowledge, the pattern selecting statistical tools have reached a high standard with minor options for further improvement. Thus, the big hurdles are defined by the problem related to cellular mixture, biological importance, and molecular networks.

Few groups have tried to address the problem of cellular mixtures 1) with mathematical modelling to identify possible components related to differential composition [21, 22] or 2) with querying data sets from different types of tissues with Boolean operators [23]. The disadvantage of this modelling is that it is underdetermined by experimental data. Furthermore, models and the comparative queries do not sufficiently define the real quantitative aspect of mixtures.

Figure 1 has illustrated the possible effect of complex cellular composition on differential expression. Figure 3 presents a scoring of differentially expressed genes

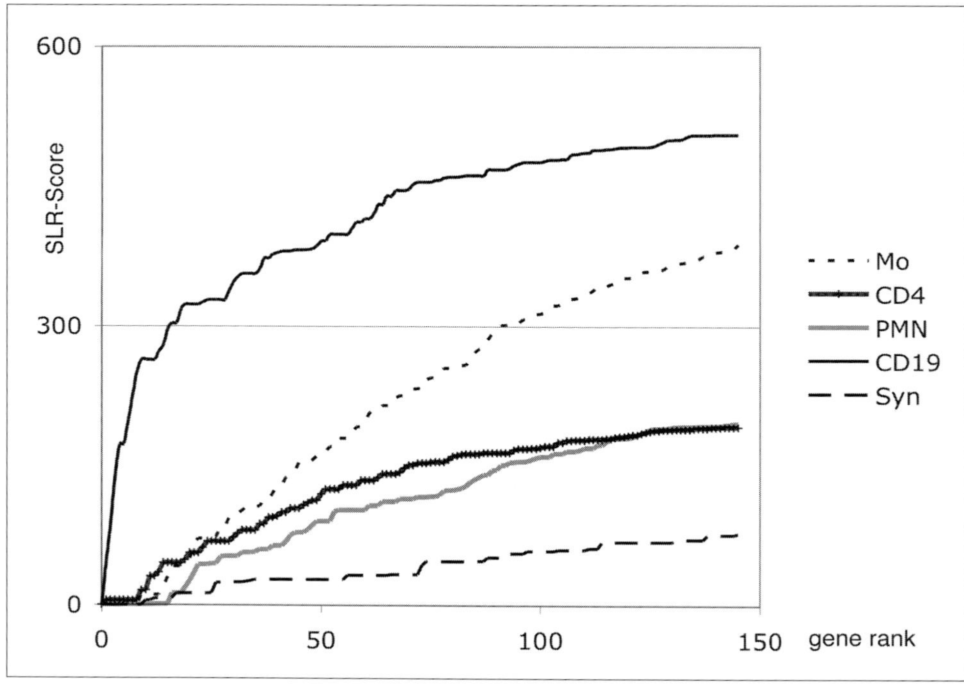

Figure 3
Scoring of the contribution of different cell types to differential gene expression in synovitis. Differentially expressed genes in RA compared to ND synovium were ranked by t-test significance (x-axis). Dominance of cell type specific expression was calculated as mean signal log ratios (SLR) by comparing normal profiles of all different cell types with each other. Cumulative scoring of the dominant SLRs (>1) of each gene was performed for each cell type (y-axis). Mo, monocytes; CD4, CD4 positive T-cells; PMN, polymorphnuclear cells; CD19, CD19 positive B-cells; Syn, normal synovial tissue.

for their cellular association, suggesting a relevant role for B-cells and monocytes in rheumatoid synovium. We have further developed this system into a basic principle to dissect complex samples into their components by comparing the profile of the complex sample to a virtual profile composed of a quantitatively comparable mixture of normal cells [24, 25]. This mixture was calculated virtually from experimentally defined signatures of highly purified cell types [26]. The composition of each sample can be determined on the basis of marker genes for each cell type. Using this approach, the analysis compares not to the control of normal tissue or normal blood but to a composed control of normal tissue and normal cell types in appropriate mixture. This model strongly depends on defined and standardised signatures. It is demanding a systematic approach but can be expanded to various experimentally determined sub-profiles of stimulated cell types. Thereby the biological importance of a gene for a given stimulus can be determined and related to a quantitative measure.

Finally, taking molecular networks into account will contribute to acknowledge the biological importance of genes with small numerical changes. To improve our still limited understanding in this context, mathematical tools are developing to calculate molecular network models on the basis of expression data [27, 28]. However, the outcome of such calculations will depend on the input of data and most of the current data sets are a confusing overlap of different cellular components which cannot be quantitatively dissected to identify the truly regulated genes. Therefore, systematic analyses will be equally important in molecular network modelling as it is for dissection of cellular mixtures. Figure 3 demonstrates that multiparameter testing may not only have to struggle with the identification of the one important gene but can gain unexpected value from the multitude of minor but concordant changes of the expression of many genes belonging to the same source. Therefore, the development of a structured systematic data collection of dependencies between genes, cells and stimuli gives a new and hopeful perspective for our goal to comprehensively analyse and interpret transcriptome data of complex biological systems.

Conclusions

Gene expression profiling provides a completely new approach to rheumatology research. As an interdisciplinary technology, it has stimulated fruitful collaboration between experts in array technology, bioinformatics, immunology and rheumatology. The molecular overview given by genome-wide profiles has revealed that many problems arise and demand systematic and structured generation of expression data. This helps to dissect the complexity of cellular mixtures in clinical samples and may also contribute to identify functional components to enable comprehensive interpretation of profiles from each patient individually. This will provide a deeper understanding in the molecular mechanisms of rheumatic diseases and advance our effort in an optimised and individualised anti-rheumatic therapy.

Acknowledgement
The experimental work which provided the basis for our expertise to write this review was supported by the German Ministry of Education and Research, grant 01GS0413.

References

1   Suzuki A, Yamada R, Chang X, Tokuhiro S, Sawada T, Suzuki M, Nagasaki M, Nakaya-ma-Hamada M, Kawaida R, Ono M et al (2003) Functional haplotypes of PADI4, encoding citrullinating enzyme peptidylarginine deiminase 4, are associated with rheumatoid arthritis. *Nat Genet* 34: 395–402

2   Heller RA, Schena M, Chai A, Shalon D, Bedilion T, Gilmore J, Woolley DE, Davis RW (1997) Discovery and analysis of inflammatory disease-related genes using cDNA microarrays. *Proc Natl Acad Sci USA* 94: 2150–2155

3   Maas K, Chan S, Parker J, Slater A, Moore J, Olsen N, Aune TM (2002) Cutting edge: molecular portrait of human autoimmune disease. *J Immunol* 169: 5–9

4   van der Pouw Kraan TC, van Gaalen FA, Kasperkovitz PV, Verbeet NL, Smeets TJ, Kraan MC, Fero M, Tak PP, Huizinga TW, Pieterman E et al (2003) Rheumatoid arthritis is a heterogeneous disease: evidence for differences in the activation of the STAT-1 pathway between rheumatoid tissues. *Arthritis Rheum* 48: 2132–2145

5   van der Pouw Kraan TC, van Gaalen FA, Huizinga TW, Pieterman E, Breedveld FC, Verweij CL (2003) Discovery of distinctive gene expression profiles in rheumatoid synovium using cDNA microarray technology: evidence for the existence of multiple pathways of tissue destruction and repair. *Genes Immun* 4: 187–196

6   Devauchelle V, Marion S, Cagnard N, Mistou S, Falgarone G, Breban M, Letourneur F, Pitaval A, Alibert O, Lucchesi C et al (2004) DNA microarray allows molecular profiling of rheumatoid arthritis and identification of pathophysiological targets. *Genes Immun* 5: 597–608

7   Morawietz L, Gehrke T, Frommelt L, Gratze P, Bosio A, Moller J, Gerstmayer B, Krenn V (2003) Differential gene expression in the periprosthetic membrane: lubricin as a new possible pathogenetic factor in prosthesis loosening. *Virchows Arch* 443: 57–66

8   Gu J, Märker-Hermann E, Baeten D, Tsai WC, Gladman D, Xiong M, Deister H, Kuipers JG, Huang F, Song YW et al (2002) A 588-gene microarray analysis of the peripheral blood mononuclear cells of spondyloarthropathy patients. *Rheumatology (Oxford)* 41: 759–766

9   Bennett L, Palucka AK, Arce E, Cantrell V, Borvak J, Banchereau J, Pascual V (2003) Interferon and granulopoiesis signatures in systemic lupus erythematosus blood. *J Exp Med* 197: 711–723

10 Baechler EC, Batliwalla FM, Karypis G, Gaffney PM, Ortmann WA, Espe KJ, Shark KB, Grande WJ, Hughes KM, Kapur V et al (2003) Interferon-inducible gene expression signature in peripheral blood cells of patients with severe lupus. *Proc Natl Acad Sci USA* 100: 2610–2615

11 Han GM, Chen SL, Shen N, Ye S, Bao CD, Gu YY (2003) Analysis of gene expression profiles in human systemic lupus erythematosus using oligonucleotide microarray. *Genes Immun* 4: 177–186

12 Crow MK, Wohlgemuth J (2003) Microarray analysis of gene expression in lupus. *Arthritis Res Ther* 5: 279–287

13 Judex M, Neumann E, Lechner S, Dietmaier W, Ballhorn W, Grifka J, Gay S, Scholmerich J, Kullmann F, Muller-Ladner U (2003) Laser-mediated microdissection facilitates analysis of area-specific gene expression in rheumatoid synovium. *Arthritis Rheum* 48: 97–102

14 Pierer M, Rethage J, Seibl R, Lauener R, Brentano F, Wagner U, Hantzschel H, Michel BA, Gay RE, Gay S et al (2004) Chemokine secretion of rheumatoid arthritis synovial fibroblasts stimulated by Toll-like receptor 2 ligands. *J Immunol* 172: 1256–1265

15 Schmutz C, Hulme A, Burman A, Salmon M, Ashton B, Buckley C, Middleton J (2005) Chemokine receptors in the rheumatoid synovium: upregulation of CXCR5. *Arthritis Res Ther* 7: R217–229

16 Michiels S, Koscielny S, Hill C (2005) Prediction of cancer outcome with microarrays: a multiple random validation strategy. *Lancet* 365: 488–492

17 Krenn V, Morawietz L, Haupl T, Neidel J, Petersen I, Konig A (2002) Grading of chronic synovitis – a histopathological grading system for molecular and diagnostic pathology. *Pathol Res Pract* 198: 317–325

18 Sickert D, Aust DE, Langer S, Haupt I, Baretton GB, Dieter P (2005) Characterization of macrophage subpopulations in colon cancer using tissue microarrays. *Histopathology* 46: 515–521

19 Shimura S, Yang G, Ebara S, Wheeler TM, Frolov A, Thompson TC (2000) Reduced infiltration of tumor-associated macrophages in human prostate cancer: association with cancer progression. *Cancer Res* 60: 5857–5861

20 Ben-Hur H, Cohen O, Schneider D, Gurevich P, Halperin R, Bala U, Mozes M, Zusman I (2002) The role of lymphocytes and macrophages in human breast tumorigenesis: an immunohistochemical and morphometric study. *Anticancer Res* 22: 1231–1238

21 Ghosh D (2004) Mixture models for assessing differential expression in complex tissues using microarray data. *Bioinformatics* 20: 1663–1669

22 Venet D, Pecasse F, Maenhaut C, Bersini H (2001) Separation of samples into their constituents using gene expression data. *Bioinformatics* 17 (Suppl 1): S279–287

23 Tureci O, Ding J, Hilton H, Bian H, Ohkawa H, Braxenthaler M, Seitz G, Raddrizzani L, Friess H, Buchler M et al (2003) Computational dissection of tissue contamination for identification of colon cancer-specific expression profiles. *Faseb J* 17: 376–385

24 Häupl T, Gruetzkau A, Gruen J, Janitz M, Stuhlmueller B, Rohrlach T, Kaps C, Rudwaleit M, Morawietz L, Zacher J et al (2004) Gene Expression Profiling in Rheumatoid Arthritis: Dissection of Inflammatory Phenotypes. *Arthritis Rheum* 51: 167

25 Häupl T, Grützkau A, Grün J, Bonin M, Stuhlmüller B, Kaps C, Rudwaleit M, Morawietz L, Gursche A, Zacher J et al (....) Dissection of expression profiles into functional components: a strategy to identify the pieces in the puzzle of systems biology. *Submitted*

26 Grützkau A, Grün J, Stuhlmueller B, Rudwaleit M, Sieper J, Gerstmayer B, Bosio A, Baumgrass R, Berek C, Humaljoki T et al (2004) Peripheral blood cell type-specific transcriptome analysis allows diagnosis and prediction of rheumatic diseases and prognosis of therapeutic impact. *Arthritis Rheum* 51 (Suppl): 1738

27 Dejori M, Stetter M (2004) Identifying interventional and pathogenic mechanisms by generative inverse modeling of gene expression profiles. *J Comput Biol* 11: 1135–1148

28 Deng X, Geng H, Ali H (2005) EXAMINE: A computational approach to reconstructing gene regulatory networks. *Biosystems* 81: 125–136

The use of humanized MHC mouse strains for studies of rheumatic diseases

Kary A. Latham[1], Timothy D. Kayes[2], Zhaohui Qian[1] and Edward F. Rosloniec[1,2,3]

Departments of [1]Medicine and [2]Pathology, University of Tennessee Health Science Center, Memphis, TN 38163, USA; [3]Veterans Affairs Medical Center, Research Service (151), 1030 Jefferson Avenue, Memphis, TN 38104, USA

Introduction

The association between the expression of specific alleles of the major histocompatibility complex (MHC) and susceptibility to a number of autoimmune diseases has been recognized for many years. After the discovery that MHC molecules bind and present antigenic peptides, there were high hopes that the antigens that drive autoimmune responses would be identified and new treatments for these diseases devised. Despite the advances made in understanding the function of MHC molecules, little new information has been generated as to how these MHC molecules participate in the pathogenesis of these diseases. This is perhaps no better exemplified than by the rheumatic diseases, most of which appear to be autoimmune in nature. While most autoimmune rheumatic diseases have clear associations with the expression of specific human leukocyte antigens (HLA) Class I or Class II alleles, it is still unclear how possessing these HLA alleles predisposes an individual to developing these diseases. In an attempt to answer these questions, a number of investigators have developed humanized mouse models of rheumatic diseases in which the HLA alleles associated with susceptibility to these diseases have been established as a transgene (see Tab. 1). In this review, we examine a number of these HLA humanized models and discuss their overall contribution to advancing our understanding of rheumatic diseases. While none of these mouse models perfectly mimics its respective human disease, it is clear that these humanized models can be valuable tools in developing new information about the function of these HLA molecules and they have provided important new insights into the pathogenesis of autoimmunity.

Humanized mice as models for rheumatoid arthritis

While rheumatoid arthritis (RA) remains an autoimmune disease of unknown etiology, susceptibility to RA is clearly associated with the expression of several human leukocyte antigens (HLA), including HLA-DR1 (DRB1*0101), HLA-DR4

The Hereditary Basis of Rheumatic Diseases, edited by Rikard Holmdahl
© 2006 Birkhäuser Verlag Basel/Switzerland

Table 1 - Humanized mouse models of rheumatic diseases

Disease	HLA gene in humanized model	Model antigen	Refs.
Lyme arthritis	DRB1*0401	CII	[74]
Rheumatoid arthritis	DRB1*0101	CII	[10]
Rheumatoid arthritis	DRB1*0401	CII	[9, 22]
Rheumatoid arthritis	DQB1*0302	CII	[40]
Rheumatoid arthritis	DRB1*0401 DQB1*0302	Proteoglycan	[33]
Rheumatoid arthritis	DRB1*0401	HCgp39	[30]
Sjögrens syndrome	DRB1*1502, DRB1*0301 DQB1*0601, DQB1*0302	RhRo60	[77]
Spondyloarthropathies	B27*2705	Spontaneous	[55, 57]
Systemic lupus erythematosus	DRB1*1502 DRB1*0301	Spontaneous	[78]

(DRB1*0401, *0404, *0405, and *0408), HLA-DR10 (DRB1*1001), and HLA-DR6 (DRB1*1402) [1–3]. The relative risk and severity of RA varies among these susceptible HLA alleles, with Caucasian individuals expressing the *0401 allele having a higher risk for developing RA and a more severe and erosive disease [1, 4], while the *0405 allele is strongly associated with RA in the Asian population [5]. Despite our current knowledge of the structure and function of Class II molecules, little is known about how the function of these RA-associated DR alleles mediates the immunopathogenesis of this disease. In addition, considering both the heterogeneity of the human population and the linkage disequilibrium that exists among HLA alleles, it has been very difficult to determine the role of individual HLA alleles and the relationship between these alleles and various potential autoantigens in RA patients. To address these questions, several research groups have developed HLA transgenic humanized mice expressing a variety of HLA-DR alleles associated with RA. The benefits of using these humanized mice as a model to study the immunopathogenesis of RA are several fold. The humanized mice have the same genetic background and can be studied in a controlled environment. In addition, since the humanized mice express only one HLA molecule, the functional role of individual HLA Class II molecules can be unambiguously assessed in studies of the presentation of candidate autoantigens as well as pathogenic mechanisms that lead to the development of autoimmune arthritis in these humanized models.

One problem that needed to be overcome in the design and use of humanized HLA Class II transgenic mice was that murine CD4, a co-receptor for Class II expressed by T-cells, does not interact with human Class II. The lack of Class II:CD4 binding can significantly alter both T-cell thymic selection and initiation of T-cell immune responses in the periphery [6]. To solve this problem, two different strategies have been applied. One is to establish double transgenic mice that express human CD4 on T-cells and HLA Class II molecules on antigen presenting cells [7]. The second approach has been to engineer chimeric HLA molecules in which the α1 and β1 domains are derived from the HLA-DR molecule but the α2 and β2 domains are derived from murine I-E [8–10]. I-E was selected because its sequence is nearly identical to DR. In both cases, the transgenic mice develop normally, and DR transgenes are fully functional in these mice both in terms of T-cell selection and the generation of T-cell immune responses [11–13].

CII autoimmunity in DRB1*0101 and *0401 transgenic mice

While several autoantigens have been proposed to be the target of the autoimmunity of RA, most of the proteins that have been studied are found predominantly in joints, including type II collagen (CII), human cartilage glycoprotein 39, and proteoglycan. CII is the major structural protein in articular cartilage, and anti-CII antibodies have been found in a high percentage of RA patients [14–18] and the presence of CII-specific T-cells has also been reported in RA patients [19–21]. In order to experimentally test the relationship between RA and CII autoimmunity, several investigators have developed HLA-DR humanized mouse models to determine if the DR1 or DR4 susceptibility alleles mediate this CII-specific autoimmunity. CII immunization of humanized, transgenic mice expressing either DRB1*0101 or DRB1*0401 on B10.M or DBA/1 backgrounds induces an inflammatory autoimmune arthritis that shares significant clinical and histopathological similarities with RA [9, 10, 22]. Surprisingly, T-cells from both DRB1*0101 and DRB1*0401 transgenic mice recognize the same dominant epitope in CII, located at amino acids 263–270 (CII263–270) [9, 10, 22]. This CII dominant epitope also binds to DRB1*0101 and DRB1*0401 in a very similar manner with the phenylalanine at amino acid 263 of the CII peptide fitting in the P1 binding pocket of both the DR1 and DR4 molecules [23]. The significance of this CII peptide is supported by recent studies in RA patients in which a T-cell response to CII263*270 was detected in patients with early disease (less than 3 years) but was less prevalent in RA patients with disease of longer duration [19].

Recently Latham et al. and Svendsen et al. demonstrated that DR1 and DR4 tetramers in which the respective peptide binding sites are loaded with the CII dominant peptide can be used to identify DR-restricted, CII-specific T-cells *ex vivo* using DR1 and DR4 humanized mouse models [12, 24]. This approach allowed for quantitation of the CII-specific T-cell response over time as arthritis develops as well as a means of isolating the pathogenic T-cell population for functional and molecular studies. The number of CII- specific T-cells reached a maximum around 10 days post immunization, and then gradually decreased, although CII-specific T-cells could still be detected at more than 100 days after immunization [12]. Cytokine analysis has revealed that these CII-specific T-cells express cytokines consistent with the Th1 phenotype [12, 24]. Moreover, these CII-specific T-cells are also present in the arthritic joints of the DR-transgenic mice but only at the early stages of disease, potentially indicating a direct role for CII-specific T-cells in the early pathogenic phase of CIA [12, 24]. DR4-CII tetramers have been also used to search for CII-specific T-cells in the synovial fluid of RA patients [25], but to date, this approach has not yielded positive results. However, based on the data from the humanized models above, the use of samples from patients with late stage disease may have contributed to the inability to detect CII-specific T-cells. Alternatively, studies using human CII-transgenic, DR-humanized mice have suggested that glycosylated forms of CII, that were not present in the tetramers, may play an important role in the autoimmunity of RA [26].

HCgp39 and DRB1*0401 transgenic mice

Human cartilage glycoprotein 39 (HCgp39) is another cartilage-derived protein that has been studied in humanized mice as a potential autoantigen in RA. HCgp39 has received attention as a potential RA antigen because HCgp39-reactive T-cells have been detected in peripheral blood from RA patients [27]. Although HCgp39 is present in many tissues, it is undetectable in healthy cartilage explants [28, 29], but is abundantly present in inflamed arthritic joints where it is produced by synovial cells and articular chondrocytes [28, 29]. Cope et al. studied the immunogenicity of HCgp39 in both transgenic mice expressing the RA-susceptibility allele DRB1*0401 as well as mice expressing a non susceptible allele, DRB1*0402 [30]. Both of these humanized models incorporated a human CD4 transgene expressed by T-cells to function as a co-receptor for the DR molecule. Interestingly, although there was no difference in T-cell proliferation between DRB1*0401 and *0402 transgenic mice following stimulation with HCgp39, the cytokine profiles of HCgp39-specific T-cells from *0401- and *0402-transgenic mice were significantly different [30]. HCgp39-specific T-cells from *0401 mice produced high levels of IFN-γ and TNF-α. In contrast, very little IFN-γ was produced by *0402-restricted T-cells, and TNF-α production was undetectable in these cells. The immunodominant epitopes of HCgp39 in *0401 and *0402 mice have also been identified, and are distinct for each allele. There are three DRB1*0401-restricted dominant epitopes, HCgp39(100–115), HCgp39(262–276), and HCgp39(322–337), while there are two major epitopes for *0402, HCgp39(22–37) and HCgp39(298–313) [30]. Additionally, proliferative responses to all three *0401-restricted immunogenic peptides have been observed in RA patients expressing the DRB1*04 susceptibility alleles

[30]. While immunization of the DRB1*0401 transgenic mice with HCgp39 induces a DR4-restricted T-cell response, these mice do not develop arthritis, in contrast to the other humanized models studying cartilage derived antigens. Despite this drawback, these mice still proved to be invaluable in testing the DR4-restricted immunogenicity of HCgp39 and identifying its antigenic determinants.

Proteoglycan and DRB1*0401 transgenic mice

Similar to CII, proteoglycan is also a major structural component of the extracellular matrix of cartilage and has been suggested to be a target of RA autoimmunity. Years ago it was demonstrated that immunization of wild type BALB/c or C3H mice with human proteoglycan induced a progressive polyarthritis that shares many similarities with RA [31, 32]. Recently, BALB/c-DRB1*0401 and -DQ8 transgenic mice were developed to study the role of proteoglycan specific, DR4- and DQ8-restricted immune responses [33]. In contrast to the humanized CII models in which introduction of DR4 and DR1 transgenes converted a CIA non-susceptible strain to a CIA susceptible strain [9, 10], proteoglycan induced arthritis (PGIA) only occurred in the humanized mice in which the transgene was expressed on the PGIA susceptible BALB/c background [33]. The incidence and severity of disease after immunization was lower compared to the standard BALB/c PGIA model, and arthritis onset occurred much later, but the histopathology was indistinguishable [33]. The transgenic mice used in this model did not express human CD4 nor was the DR molecule chimeric for I-E, therefore the lack of a CD4 co-receptor for the DR molecule may have contributed to the reduced arthritogenicity of the proteoglycan in these transgenic mice. Nevertheless, these humanized models allowed the investigators to identify antigenic components of human proteoglycan that are bound and presented by these RA alleles, and studies of RA patients with these proteoglycan peptides have indicated that T-cell immunity to these peptides exists in some patients [34].

DQ8 transgenic mice and RA

As an alternate hypothesis to the role of the HLA-DR in mediating susceptibility to RA, Zanelli et al. proposed that HLA-DQ8, and not DR1 or DR4, directly mediates the immunopathogenesis in RA [35]. Although the role of DQ8 in RA is still in dispute [36, 37], human population studies have indicated that there is linkage disequilibrium between HLA-DQ8 and DR4 [38], suggesting a potential role of DQ8 in RA. To address this possibility, DBA/1-DQ8 transgenic mice in which murine Class II genes were deleted were generated for studying the function of DQ8 [39]. Immunization of these DQ8 transgenic mice with CII induced an autoimmune arthritis similar to the other CIA models, and immunodominant epitopes of CII presented by

DQ8 in these mice were also identified. In contrast to the DR1- or DR4-restricted T-cell responses to CII in which only two determinants were identified, the DQ8 mice generated T-cell responses to 15 immunogenic epitopes of CII [40] and none of them overlapped with the DR-specific CII epitopes. While these data clearly indicate the differences between DR and DQ in specificity of peptide binding, perhaps the most interesting aspect of the DQ hypothesis is its incorporation of a functional role for the shared epitope of DR alleles associated with RA susceptibility. Sequence analysis of RA-associated HLA-DRB1 alleles revealed that these alleles share a common motif, a 'shared epitope' at residues 70–74, that is not present in other DR alleles [41]. The amino acid sequence for the shared epitope is Q/R-K/R-R-A-A, and at least a few of these residues of the shared epitope are involved in peptide binding within the P4 binding pocket of the DR molecules [42]. However, what exact role the shared epitope plays in RA has remained a mystery, although a number of hypotheses have been proposed. [43–48]. Zanelli et al. have proposed that the polymorphic region of DRB1 that encodes the shared epitope is a source of an antigenic peptide that regulates the function of the DQ8 molecule by binding to the DQ peptide binding site [35]. This hypothesis is supported by their own studies in the DQ8 mice indicating that peptides from this region derived from DR alleles not associated with RA generate DQ8-restricted T-cell responses, while peptides from RA alleles do not [49]. Their conclusion is that the peptide from non-RA alleles binds to DQ8 and regulates its function, whereas the analogous peptide derived from RA alleles lacks this ability. These data were supported by the fact that the susceptibility of DQ8 transgenic mice to CII immunization can also be blocked by introducing an RA-resistant DRB1*0402 transgene. The presence of this non-susceptibility allele also switches the cytokine profile of T-cells responding to CII immunization to the Th2 phenotype [50]. Whether or not this intriguing hypothesis is correct, these studies demonstrate the utility of humanized mouse strains in testing novel hypotheses for the immunopathogenesis of RA.

HLA-B27 transgenic mice as models for spondyloarthropathy

Among the strongest associations between HLA haplotype and rheumatic disease is the correlation between expression of the MHC Class I allele HLA-B27 and the incidence of spondyloarthropathy. The spondyloarthropathies are a group of related diseases typified by sacroiliitis, peripheral arthritis, and enthesitis, or inflammation at the site of the attachment of skeletal muscle, tendons, ligaments, or joint capsule to bone (for a review of the spondyloarthropathies see Kataria et al., 2004) [51]. Prominent among the spondyloarthritides is ankylosing spondylitis (AS), a chronic disease involving progressive ossification of the vertebrae and a host of other symptoms, including arthritis in the hips and shoulders, enthesitis, and acute uveitis. AS affects as much as 0.5% of the population with 96% of affected individuals expressing HLA-B27 [52]. A second spondyloarthropathy showing high correlation with HLA-B27 expression is reactive arthritis (ReA), an acute disease that is precipitated by infection in the genitourinary or gastrointestinal tracts and is typified by joint inflammation, enthesitis, and uveitis. ReA affects only 0.1% of the population yet 80% of reactive arthritis patients express HLA-B27.

Murine models for spondyloarthropathy

Several rodent models have been developed for studying the role of HLA-B27 in conferring susceptibility to rheumatic diseases. Mice and rats expressing HLA-B27 as a transgene develop arthritic symptoms, including swelling of peripheral joints and nail deformation, which vary according to species, the conditions in which the animal was raised, the copy number of the transgene, and the inclusion of additional genetic modifications. The HLA-B27 transgenic model was first established in rats, which develop spontaneous arthritic symptoms when expression of HLA-B27 is paired with expression of human β_2-microglobulin (β_2m) [53]. Disease in this model is also dependent upon a high copy number of HLA-B27, and upon raising the animals in an environment that is not germ-free [54]. Mice transgenic only for HLA-B27 do not develop disease. Spontaneous arthritis in HLA-B27 transgenic mice requires a deficiency in murine β_2m and housing conditions that are not germ-free [55, 56] or the inclusion of human β_2m as a second transgene [57]. Removal from germ-free colonies is also essential to arthritis induction in the mice transgenic for both HLA-B27 and human β_2m. The lack of requirement for β_2m in these mice is interesting, since normal functions of MHC Class I require pairing with β_2m prior to peptide binding and presentation of peptide by MHC Class I on the cell surface.

The HLA-B27 molecule may be prone to defective function

There are several theories that have been developed in order to try to explain how the HLA-B27 molecule mediates arthritis in mice and rats, and how β_2m plays a role in the pathogenesis. Among the first theories presented was that the HLA-B27 molecule itself was a target of the immune system. Khare et al. found that whereas HLA-B27 was not expressed on the surface of cells in HLA-B27$^+$/ β_2m$^{-/-}$ mice, expression could be induced by stimulation of cells from these mice with concanavalin A (con A) [55]. This cell surface expression of HLA-B27 in the absence of β_2m was hypothesized by Khare et al. to result in an immune response against the HLA-B27$^+$ cells by CD8$^+$ T-cells that had not been exposed to HLA-B27 during their thymic development. Alternatively, Mear et al. have suggested that the amino acid composition of the peptide-binding B pocket of HLA-B27 [58] causes the HLA-B27 molecule to fold inefficiently [59]. These investigators hypothesized that this misfolding, which normally triggers protein degradation, could result in an excess of intracellular HLA-B27 in the protein degradation pathways. This could result in either an excess of HLA-B27-derived peptides being presented by other Class I molecules, or possibly the B27-derived peptides being shunted into the Class II presentation pathway. The potential involvement of CD4+ T-cells in HLA-B27-associated pathogenesis is an intriguing possibility that will be discussed later in this section.

The concept that excess HLA-B27 contributes to autoimmunity was also tested by Allen et al. who found that HLA-B27 could form homodimers via an unpaired cysteine residue in the molecule's $\alpha 1$ domain [60]. These homodimers were expressed on the cell surface in the absence of β_2m and retained some peptide binding ability. This finding provided a possible mechanism for the spontaneous arthritis occurring in HLA-B27 transgenic mice deficient in β_2m, i.e., that imperfectly folded HLA-B27 that escapes degradation may be expressed on the cell surface without β_2m and either serve as an antigen itself, present peptides that are arthritogenic, or serve as a ligand for receptors other than T-cell receptors. A potential role of HLA-B27 homodimers (B27$_2$) is supported by the demonstration of their existence on the surface of splenocytes from HLA-B27 transgenic mice that are either lacking β_2m or are also transgenic for human β_2m [61]. To determine if these homodimers bound a ligand other than T-cell receptor, Kollnberger et al. used tetrameric B27$_2$ to probe murine splenocytes. They found that the B27$_2$ tetramer bound to a subset of splenocytes via the paired Ig-like receptors, PIRs, expressed on the B-cells and monocytes in these mice. Related studies strengthened the relevance of this model to human spondyloarthropathies by showing that human AS patients express HLA-B27 homodimers as well as non T-cell receptor ligands for these homodimers [62].

HLA-B27 may present arthritogenic peptides

In addition to serving as a direct mediator of disease via inefficient folding and inappropriate dimerization, it has also been suggested that HLA-B27 may precipitate spondyloarthropathy by presenting arthritogenic peptides. The fact that the environment in which mice transgenic for HLA-B27 are raised affects arthritis incidence suggests that presentation of bacterial antigens may play a role in HLA-B27-mediated disease. Most of the rodent-based experiments involving bacterial peptides and HLA-B27 have been performed using an HLA-B27 transgenic rat model that is dependent upon β_2m expression. Using LEW rats transgenic for HLA-B27 and human β_2m, Popov et al. found that when the rats were immunized against HLA-B27 and restimulated *in vitro* with *Chlamydia* infected B27+ cells, CTL developed to both B27+ targets as well as *Chlamydia* infected syngeneic targets [63, 64]. These data indicated that CTLs were generated that recognize both HLA-B27 and

Chlamydia, implicating "a dynamic relationship between recognition of B27 peptides and *Chlamydia* peptides", although evidence of cross recognition of peptides is still lacking [63].

HLA-B27 animal models have also been useful for the study of other candidate antigens for AS, including autologous proteins such as the cartilage protein aggrecan. Using HLA-B27 transgenic mice immunized with human aggrecan peptides, Kuon et al. found that CD8+ T-cells from these mice were stimulated only by the human peptides and not by their murine counterparts [65]. The relevance of this murine HLA-B27 transgenic model is supported by a clinical study performed by Atagunduz et al. who found that CD8+ T-cells from synovial fluid and blood mononuclear cells from HLA-B27+ AS patients could be stimulated by cartilage antigens [66].

HLA-B27 may stimulate CD4+ T-cells

The expression of HLA-B27 on the cell surface in the absence of β_2m suggests that the pathogenicity of HLA-B27 expression may involve a mechanism other than the normal interaction between peptide-presenting MHC Class I molecules and CD8+ T-cells. There is evidence in human AS patients of the presence of CD4+ T-cells capable of interacting with HLA-B27 [67]. Roddis et al. investigated this phenomenon in the mouse model via the creation of mice transgenic both for a human T-cell receptor and for HLA-B27 [68]. They found that CD4+ T-cells expressing the transgenic receptor were capable of functioning as HLA-B27-restricted T-cells and as such were activated and able to produce Th1 or Th2 cytokines. In all, the experiments performed with HLA-B27-transgenic mice suggest that the pathogenic effect of HLA-B27 may be achieved through a variety of mechanisms, ranging from inappropriate cell-surface expression of the molecule, leading to recognition of HLA-B27 by the body's own immune system, to presentation of pathogenic peptides in the context of normal β_2m -paired function. These humanized murine models have been essential in elucidating these pathogenic mechanisms and thereby providing direction for future immune-based therapies for rheumatic diseases linked to HLA-B27 expression.

Murine models for Lyme arthritis

Autoimmune arthritis associated with Lyme disease is yet another example where susceptibility to autoimmunity is strongly associated with expression of a specific HLA-DR allele. However, unlike almost all other autoimmune diseases, a causative agent for this autoimmune disease is known. Lyme disease is caused by an infection with *Borrelia burgdorferi* and the autoimmune arthritis develops as a sequela to this infection. Approximately 10% of infected individuals develop a chronic joint inflammation that resists treatment with antibiotics [69]. This arthritis persists long after *B. burgdorferi* is detectable in the patient and is believed to be an autoimmune response triggered by molecular mimicry involving a self protein, human leukocyte function-associated antigen 1 (LFA-1), and a bacterial protein, outer surface protein A (OspA), expressed by the infecting organism [70]. Treatment-resistant Lyme arthritis that persists for an extended duration (12–48 months) is strongly associated with the expression of either HLA-DR4 or HLA-DR2 [71]. 57% of the patients that develop Lyme arthritis express HLA-DR4 (DRB1*0401), whereas 75% of the non-HLA-DR4 patients express HLA-DR2.

Humanized mouse models have been instrumental in advancing our understanding of the mechanism of autoimmunity associated with Lyme arthritis. Mice expressing HLA-DRB1*0401 as a transgene have been used to identify and characterize the antigenic peptides thought to drive the development of autoimmune arthritis in Lyme disease. In 1994, Steere et al. identified the OspA protein from *B. burgdorferi* as a primary immunogenic protein that contains an antigenic peptide that binds to HLA-DRB1*0401, and is recognized by T-cell lines developed from patients with treatment-resistant arthritis [69, 70, 72, 73]. Using HLA-DRB1*0401 transgenic mice for *in vivo* studies, these investigators were able to identify specific epitopes of OspA that stimulated CD4$^+$ T-cells restricted to this DR4 subtype. Of the peptides identified, the OspA(165–173) peptide bound with the highest affinity to DR4. While the identification of this antigenic peptide was informative, how an immune response to this peptide or the OspA protein led to the development of autoimmune arthritis was still unknown. Steere and colleagues proposed that molecular mimicry between the OspA protein and a human protein was the mechanism by which autoimmunity developed, and identified a human protein, LFA-1, that contained a similar peptide sequence, LFA-1(326–343). When tested with T-cells from the synovial fluid of patients with treatment-resistant Lyme arthritis, this LFA-1 peptide was found to stimulate these T-cells [74]. The association between OspA and LFA-1 was confirmed by Trollmo et al. who discovered CD4$^+$ T-cells reactive for both OspA and LFA-1 peptides in both OspA-immunized HLA-DR4 transgenic mice and in human Lyme arthritis patients [75].

One drawback of this humanized mouse model is that these mice are not susceptible to autoimmune arthritis when they are infected with *B. burgdorferi* [76]. The obvious explanation for this, based on the molecular mimicry data, is that the sequence of the mouse LFA-1 protein is different from the human LFA-1 protein within the 326–343 region, and this is indeed the case [74]. It would be interesting to see this model of molecular mimicry tested by the production of a mouse model expressing both the DR4 molecule and human LFA-1. If such a mouse model were susceptible to OspA-induced autoimmune arthritis, these data would provide strong evidence of a role for molecular mimicry in *B. burgdorferi*-mediated autoimmunity.

References

1   Nepom GT, Byers P, Seyfried C, Healey LA, Wilske KR, Stage D, Nepom BS (1989) HLA genes associated with rheumatoid arthritis. Identification of susceptibility alleles using specific oligonucleotide probes. *Arthritis Rheum* 32: 15–21

2   Wordsworth BP, Lanchbury JS, Sakkas LI, Welsh KI, Panayi GS, Bell JI (1989) HLA-DR4 subtype frequencies in rheumatoid arthritis indicate that DRB1 is the major susceptibility locus within the HLA class II region. *Proc Natl Acad Sci USA* 86: 10049–10053

3   Ferucci ED, Templin DW, Lanier AP (2005) Rheumatoid arthritis in American Indians and Alaska Natives: a review of the literature. *Semin Arthritis Rheum* 34: 662–667

4   Weyand CM, McCarthy TG, Goronzy JJ (1995) Correlation between disease phenotype and genetic heterogeneity in rheumatoid arthritis. *J Clin Invest* 95: 2120–2126

5   Lee HS, Lee KW, Song GG, Kim HA, Kim SY, Bae SC (2004) Increased susceptibility to rheumatoid arthritis in Koreans heterozygous for HLA-DRB1*0405 and *0901. *Arthritis Rheum* 50: 3468–3475

6   Law YM, Yeung RS, Mamalaki C, Kioussis D, Mak TW, Flavell RA (1994) Human CD4 restores normal T cell development and function in mice deficient in murine CD4. *J Exp Med* 179: 1233–1242

7   Fugger L, Michie SA, Rulifson I, Lock CB, McDevitt GS (1994) Expression of HLA-DR4 and human CD4 transgenes in mice determines the variable region beta-chain T-cell repertoire and mediates an HLA-DR-restricted immune response. *Proc Natl Acad Sci USA* 91: 6151–6155

8   Woods A, Chen HY, Trumbauer ME, Sirotina A, Cummings R, Zaller DM (1994) Human major histocompatibility complex class II-restricted T cell responses in transgenic mice. *J Exp Med* 180: 173–181

9   Rosloniec EF, Brand DD, Myers LK, Esaki Y, Whittington KB, Zaller DM, Woods A, Stuart JM, Kang AH (1998) Induction of autoimmune arthritis in HLA-DR4 (DRB1*0401) transgenic mice by immunization with human and bovine type II collagen. *J Immunol* 160: 2573–2578

10 Rosloniec EF, Brand DD, Myers LK, Whittington KB, Gumanovskaya M, Zaller DM, Woods A, Altmann DM, Stuart JM, Kang AH (1997) An HLA-DR1 transgene confers susceptibility to collagen-induced arthritis elicited with human type II collagen. *J Exp Med* 185: 1113–1122

11 He X, Rosloniec EF, Myers LK, McColgan WL, 3rd, Gumanovskaya M, Kang AH, Stuart JM (2004) T cell receptors recognizing type II collagen in HLA-DR-transgenic mice characterized by highly restricted V beta usage. *Arthritis Rheum* 50: 1996–2004

12 Latham KA, Whittington KB, Zhou R, Qian Z, Rosloniec EF (2005) *Ex vivo* characterization of the autoimmune T cell response in the HLA-DR1 mouse model of collagen-induced arthritis reveals long-term activation of Type II collagen-specific cells and their presence in arthritic joints. *J Immunol* 174: 3978–3985

13 Fugger L, Rothbard JB, Sonderstrup-McDevitt G (1996) Specificity of an HLA-DRB1*0401-restricted T cell response to type II collagen. *Eur J Immunol* 26: 928–933

14 Kim WU, Yoo WH, Park W, Kang YM, Kim SI, Park JH, Lee SS, Joo YS, Min JK, Hong YS et al (2000) IgG antibodies to type II collagen reflect inflammatory activity in patients with rheumatoid arthritis. *J Rheumatol* 27: 575–581

15 Michaeli D, Fudenburg HH (1974) The incidence and antigenic specificity of antibodies against denatured human collagen in rheumatoid arthritis. Clin Immunol Immunolpathol 2: 153–159

16 Stuart JM, Huffstutter EH, Townes AS, Kang AH (1983) Incidence and specificity of antibodies to type I, II, III, IV, and V collagen in rheumatoid arthiritis and other rheumatic diseases as measured by 125I-radioimmunoassay. *Arthritis Rheum.* 26: 832–840

17 Terato K, Shimozuru Y, Katayama K, Takemitsu Y, Yamashita I, Miyatsu M, Fujii K, Sagara M, Kobayashi S, Goto M et al (1990) Specificity of antibodies to type II collagen in rheumatoid arthritis. *Arthritis Rheum* 33: 1493–1500

18 Watson W, Cremer M, Wooley P, Townes A (1986) Assessment of the potential pathogenicity of type II collagen autoantibodies in patients with rheumatoid arthritis. Arth Rheum 29: 1316–1321

19 Kim HY, Kim WU, Cho ML, Lee SK, Youn J, Kim SI, Yoo WH, Park JH, Min JK, Lee SH et al (1999) Enhanced T cell proliferative response to type II collagen and synthetic peptide CII (255-274) in patients with rheumatoid arthritis. *Arthritis Rheum* 42: 2085–2093

20 Londei M, Savill CM, Verhoef A, Brennan F, Leech ZA, Duance V, Maini RN, Feldman M (1989) Persistence of collagen type II-specific T-cell clones in the synovial membrane of a patient with rheumatoid arthritis. *Proc Natl Acad Sci USA* 86: 636–640

21 He X, Kang AH, Stuart JM (2000) Accumulation of T cells reactive to type II collagen in synovial fluid of patients with rheumatoid arthritis. *J Rheumatol* 27: 589–593

22 Andersson EC, Hansen BE, Jacobsen H, Madsen LS, Andersen CB, Engberg J, Rothbard JB, McDevitt GS, Malmstrom V, Holmdahl R et al (1998) Definition of MHC and T cell receptor contacts in the HLA-DR4restricted immunodominant epitope in type II collagen and characterization of collagen-induced arthritis in HLA-DR4 and human CD4 transgenic mice. *Proc Natl Acad Sci USA* 95: 7574–7579

23 Rosloniec EF, Whittington KB, Zaller DM, Kang AH (2002) HLA-DR1 (DRB1*0101) and DR4 (DRB1*0401) use the same anchor residues for binding an immunodominant peptide derived from human type II collagen. *J Immunol* 168: 253–259

24 Svendsen P, Andersen CB, Willcox N, Coyle AJ, Holmdahl R, Kamradt T, Fugger L (2004) Tracking of proinflammatory collagen-specific T cells in early and late collagen-induced arthritis in humanized mice. *J Immunol* 173: 7037–7045

25 Kotzin BL, Falta MT, Crawford F, Rosloniec EF, Bill J, Marrack P, Kappler J (2000) Use of soluble peptide-DR4 tetramers to detect synovial T cells specific for cartilage antigens in patients with rheumatoid arthritis. *Proc Natl Acad Sci USA* 97: 291–296

26 Backlund J, Carlsen S, Hoger T, Holm B, Fugger L, Kihlberg J, Burkhardt H, Holmdahl

R (2002) Predominant selection of T cells specific for the glycosylated collagen type II epitope (263-270) in humanized transgenic mice and in rheumatoid arthritis. *Proc Natl Acad Sci USA* 99: 9960–9965

27 Verheijden GF, Rijnders AW, Bos E, Coenen-de Roo CJ, van Staveren CJ, Miltenburg AM, Meijerink JH, Elewaut D, de Keyser F, Veys E et al (1997) Human cartilage glycoprotein-39 as a candidate autoantigen in rheumatoid arthritis. *Arthritis Rheum* 40: 1115–1125

28 Hakala BE, White C, Recklies AD (1993) Human cartilage gp-39, a major secretory product of articular chondrocytes and synovial cells, is a mammalian member of a chitinase protein family. *J Biol Chem* 268: 25803–25810

29 Johansen JS, Jensen HS, Price PA (1993) A new biochemical marker for joint injury. Analysis of YKL-40 in serum and synovial fluid. *Br J Rheumatol* 32: 949–955

30 Cope AP, Patel SD, Hall F, Congia M, Hubers HA, Verheijden GF, Boots AM, Menon R, Trucco M, Rijnders AW et al (1999) T cell responses to a human cartilage autoantigen in the context of rheumatoid arthritis-associated and nonassociated HLA-DR4 alleles. *Arthritis Rheum* 42: 1497–1507

31 Glant TT, Bardos T, Vermes C, Chandrasekaran R, Valdez JC, Otto JM, Gerard D, Velins S, Lovasz G, Zhang J et al (2001) Variations in susceptibility to proteoglycan-induced arthritis and spondylitis among C3H substrains of mice: evidence of genetically acquired resistance to autoimmune disease. *Arthritis Rheum* 44: 682–692

32 Glant TT, Mikecz K, Arzoumanian A, Poole AR (1987) Proteoglycan-induced arthritis in BALB/c mice. Clinical features and histopathology. *Arthritis Rheum* 30: 201–212

33 Szanto S, Bardos T, Szabo Z, David CS, Buzas EI, Mikecz K, Glant TT (2004) Induction of arthritis in HLA-DR4-humanized and HLA-DQ8-humanized mice by human cartilage proteoglycan aggrecan but only in the presence of an appropriate (non-MHC) genetic background. *Arthritis Rheum* 50: 1984–1995

34 Guerassimov A, Zhang Y, Banerjee S, Cartman A, Leroux JY, Rosenberg LC, Esdaile J, Fitzcharles MA, Poole AR (1998) Cellular immunity to the G1 domain of cartilage proteoglycan aggrecan is enhanced in patients with rheumatoid arthritis but only after removal of keratan sulfate. *Arthritis Rheum* 41: 1019–1025

35 Zanelli E, Gonzalez-Gay MA, David CS (1995) Could HLA-DRB1 be the protective locus in rheumatoid arthritis? *Immunol Today* 16: 274–278

36 Fugger L, Svejgaard A (2000) Association of MHC and rheumatoid arthritis. HLA-DR4 and rheumatoid arthritis: studies in mice and men. *Arthritis Res* 2: 208–211

37 Taneja V, David CS (2000) Association of MHC and rheumatoid arthritis. Regulatory role of HLA class II molecules in animal models of RA: studies on transgenic/knockout mice. *Arthritis Res* 2: 205–207

38 Ilonen J, Reijonen H, Arvilommi H, Jokinen I, Mottonen T, Hannonen P (1990) HLA-DR antigens and HLA-DQ beta chain polymorphism in susceptibility to rheumatoid arthritis. *Ann Rheum Dis* 49: 494–496

39 Zhou P, Anderson GD, Savarirayan S, Inoko H, David CS (1991) Human HLA-DQ beta

chain presents minor lymphocyte stimulating locus gene products and clonally deletes TCR V beta 6+, V beta 8.1+ T cells in single transgenic mice. *Hum Immunol* 31: 47–56

40 Krco CJ, Watanabe S, Harders J, Griffths MM, Luthra H, David CS (1999) Identification of T cell determinants on human type II collagen recognized by HLA-DQ8 and HLA-DQ6 transgenic mice. *J Immunol* 163: 1661–1665

41 Gregersen PK, Silver J, Winchester RJ (1987) The shared epitope hypothesis. An approach to understanding the molecular genetics of susceptibility to rheumatoid arthritis. *Arthritis Rheum* 30: 1205–1213

42 Dessen A, Lawrence CM, Cupo S, Zaller DM, Wiley DC (1997) X-ray crystal structure of HLA-DR4 (DRA*0101, DRB1*0401) complexed with a peptide from human collagen II. *Immunity* 7: 473–481

43 Hammer J, Gallazzi F, Bono E, Karr RW, Guenot J, Valsasnini P, Nagy ZA, Sinigaglia F (1995) Peptide binding specificity of HLA-DR4 molecules: correlation with rheumatoid arthritis association. *J Exp Med* 181: 1847–1855

44 Wucherpfennig KW, Strominger JL (1995) Selective binding of self peptides to disease-associated major histocompatibility complex (MHC) molecules: a mechanism for MHC-linked susceptibility to human autoimmune diseases. *J Exp Med* 181: 1597–1601

45 Walser-Kuntz DR, Weyand CM, Weaver AJ, O'Fallon WM, Goronzy JJ (1995) Mechanisms underlying the formation of the T cell receptor repertoire in rheumatoid arthritis. *Immunity* 2: 597–605

46 La Cava A, Nelson JL, Ollier WE, MacGregor A, Keystone EC, Thorne JC, Scavulli JF, Berry CC, Carson DA, Albani S (1997) Genetic bias in immune responses to a cassette shared by different microorganisms in patients with rheumatoid arthritis. *J Clin Invest* 100: 658–663

47 Auger I, Escola JM, Gorvel JP, Roudier J (1996) HLA-DR4 and HLA-DR10 motifs that carry susceptibility to rheumatoid arthritis bind 70-kD heat shock proteins. *Nat Med* 2: 306–310

48 Auger I, Lepecuchel L, Roudier J (2002) Interaction between heat-shock protein 73 and HLA-DRB1 alleles associated or not with rheumatoid arthritis. *Arthritis Rheum* 46: 929–933

49 Zanelli E, Krco CJ, Baisch JM, Cheng S, David CS (1996) Immune response of HLA-DQ8 transgenic mice to peptides from the third hypervariable region of HLA-DRB1 correlates with predisposition to rheumatoid arthritis. *Proc Natl Acad Sci USA* 93: 1814–1819

50 Taneja V, Taneja N, Behrens M, Pan S, Trejo T, Griffiths M, Luthra H, David CS (2003) HLA-DRB1*0402 (DW10) transgene protects collagen-induced arthritis-susceptible H2Aq and DRB1*0401 (DW4) transgenic mice from arthritis. *J Immunol* 171: 4431–4438

51 Kataria RK, Brent LH (2004) Spondyloarthropathies. *Am Fam Physician* 69: 2853–2860

52 Brewerton DA, Hart FD, Nicholls A, Caffrey M, James DC, Sturrock RD (1973) Ankylosing spondylitis and HL-A 27. *Lancet* 1: 904–907

53   Hammer RE, Maika SD, Richardson JA, Tang JP, Taurog JD (1990) Spontaneous inflammatory disease in transgenic rats expressing HLA-B27 and human beta 2m: an animal model of HLA-B27-associated human disorders. *Cell* 63: 1099–1112

54   Taurog JD, Richardson JA, Croft JT, Simmons WA, Zhou M, Fernandez-Sueiro JL, Balish E, Hammer RE (1994) The germfree state prevents development of gut and joint inflammatory disease in HLA-B27 transgenic rats. *J Exp Med* 180: 2359–2364

55   Khare SD, Luthra HS, David CS (1995) Spontaneous inflammatory arthritis in HLA-B27 transgenic mice lacking beta 2-microglobulin: a model of human spondyloarthropathies. *J Exp Med* 182: 1153–1158

56   Nickerson CL, Hanson J, David CS (1990) Expression of HLA-B27 in transgenic mice is dependent on the mouse H-2D genes. *J Exp Med* 172: 1255–1261

57   Khare SD, Hansen J, Luthra HS, David CS (1996) HLA-B27 heavy chains contribute to spontaneous inflammatory disease in B27/human beta2-microglobulin (beta2m) double transgenic mice with disrupted mouse beta2m. *J Clin Invest* 98: 2746–2755

58   Colbert RA, Rowland-Jones SL, McMichael AJ, Frelinger JA (1993) Allele-specific B pocket transplant in class I major histocompatibility complex protein changes requirement for anchor residue at P2 of peptide. *Proc Natl Acad Sci USA* 90: 6879–6883

59   Mear JP, Schreiber KL, Munz C, Zhu X, Stevanovic S, Rammensee HG, Rowland-Jones SL, Colbert RA (1999) Misfolding of HLA-B27 as a result of its B pocket suggests a novel mechanism for its role in susceptibility to spondyloarthropathies. *J Immunol* 163: 6665–6670

60   Allen RL, O'Callaghan CA, McMichael AJ, Bowness P (1999) Cutting edge: HLA-B27 can form a novel beta 2-microglobulin-free heavy chain homodimer structure. *J Immunol* 162: 5045–5048

61   Kollnberger S, Bird LA, Roddis M, Hacquard-Bouder C, Kubagawa H, Bodmer HC, Breban M, McMichael AJ, Bowness P (2004) HLA-B27 heavy chain homodimers are expressed in HLA-B27 transgenic rodent models of spondyloarthritis and are ligands for paired Ig-like receptors. *J Immunol* 173: 1699–1710

62   Kollnberger S, Bird L, Sun MY, Retiere C, Braud VM, McMichael A, Bowness P (2002) Cell-surface expression and immune receptor recognition of HLA-B27 homodimers. *Arthritis Rheum* 46: 2972–2982

63   Popov I, Dela Cruz CS, Barber BH, Chiu B, Inman RD (2001) The effect of an anti-HLA-B27 immune response on CTL recognition of *Chlamydia*. *J Immunol* 167: 3375–3382

64   Popov I, Dela Cruz CS, Barber BH, Chiu B, Inman RD (2002) Breakdown of CTL tolerance to self HLA-B*2705 induced by exposure to *Chlamydia trachomatis*. *J Immunol* 169: 4033–4038

65   Kuon W, Kuhne M, Busch DH, Atagunduz P, Seipel M, Wu P, Morawietz L, Fernahl G, Appel H, Weiss EH et al (2004) Identification of novel human aggrecan T cell epitopes in HLA-B27 transgenic mice associated with spondyloarthropathy. *J Immunol* 173: 4859–4866

66   Atagunduz P, Appel H, Kuon W, Wu P, Thiel A, Kloetzel PM, Sieper J (2005) HLA-B27-

restricted CD8[+] T cell response to cartilage-derived self peptides in ankylosing spondylitis. *Arthritis Rheum* 52: 892–901

67 Boyle LH, Goodall JC, Opat SS, Gaston JS (2001) The recognition of HLA-B27 by human CD4(+) T lymphocytes. *J Immunol* 167: 2619–2624

68 Roddis M, Carter RW, Sun MY, Weissensteiner T, McMichael AJ, Bowness P, Bodmer HC (2004) Fully functional HLA B27-restricted CD4[+] as well as CD8[+] T cell responses in TCR transgenic mice. *J Immunol* 172: 155–161

69 Guerau-de-Arellano M, Huber BT (2002) Development of autoimmunity in Lyme arthritis. *Curr Opin Rheumatol* 14: 388–393

70 Dickman S (1998) Possible cause found for Lyme arthritis. *Science* 281: 631–632

71 Steere AC, Dwyer E, Winchester R (1990) Association of chronic Lyme arthritis with HLA-DR4 and HLA-DR2 alleles. *N Engl J Med* 323: 219–223

72 Lengl-Janssen B, Strauss AF, Steere AC, Kamradt T (1994) The T helper cell response in Lyme arthritis: differential recognition of Borrelia burgdorferi outer surface protein A in patients with treatment-resistant or treatment-responsive Lyme arthritis. *J Exp Med* 180: 2069–2078

73 Steere AC, Falk B, Drouin EE, Baxter-Lowe LA, Hammer J, Nepom GT (2003) Binding of outer surface protein A and human lymphocyte function-associated antigen 1 peptides to HLA-DR molecules associated with antibiotic treatment-resistant Lyme arthritis. *Arthritis Rheum* 48: 534–540

74 Gross DM, Forsthuber T, Tary-Lehmann M, Etling C, Ito K, Nagy ZA, Field JA, Steere AC, Huber BT (1998) Identification of LFA-1 as a candidate autoantigen in treatment-resistant Lyme arthritis. *Science* 281: 703–706

75 Trollmo C, Meyer AL, Steere AC, Hafler DA, Huber BT (2001) Molecular mimicry in Lyme arthritis demonstrated at the single cell level: LFA-1 alpha L is a partial agonist for outer surface protein A-reactive T cells. *J Immunol* 166: 5286–5291

76 Feng S, Barthold SW, Bockenstedt LK, Zaller DM, Fikrig E (1995) Lyme disease in human DR4Dw4-transgenic mice. *J Infect Dis* 172: 286–289

77 Paisansinsup T, Deshmukh US, Chowdhary VR, Luthra HS, Fu SM, David CS (2002) HLA class II influences the immune response and antibody diversification to Ro60/Sjogren's syndrome-A: heightened antibody responses and epitope spreading in mice expressing HLA-DR molecules. *J Immunol* 168: 5876–5884

78 Paisansinsup T, Vallejo AN, Luthra H, David CS (2001) HLA-DR modulates autoantibody repertoire, but not mortality, in a humanized mouse model of systemic lupus erythematosus. *J Immunol* 167: 4083–4090

SKG mice, a monogenic model of autoimmune arthritis due to altered signal transduction in T-cells

Shimon Sakaguchi[1,2], Takeshi Takahashi[1], Hiroshi Hata[1], Hiroyuki Yoshitomi[1], Satoshi Tanaka[1], Keiji Hirota[1], Takashi Nomura[1] and Noriko Sakaguchi[1]

[1]Department of Experimental Pathology, Institute for Frontier Medical Sciences, Kyoto University, 53 Shogoin Kawahara-cho, Sakyo-ku, Kyoto 606-8507, Japan; [2]Core Research for Evolutional Science and Technology (CREST), Science and Technology Agency of Japan, Kawaguchi, Japan

Introduction

Rheumatoid arthritis (RA) is a chronic systemic inflammatory disease of unknown etiology that primarily affects the synovial membranes of multiple joints and frequently accompanies extra-articular lesions [1, 2]. Both genetic and environmental factors contribute to the development of RA [1, 2]. Recent studies have facilitated our understanding of the process of chronic joint inflammation, in particular the roles of proinflammatory cytokines (such as TNF-α, IL-1, and IL-6) [3]. Neutralization of these cytokines or blockade of their action can indeed halt the progression of the disease [4, 5]. It is still obscure, however, how RA is triggered in the first place. RA has been suspected to be autoimmune in etiology because of the presence of autoantibodies, such as rheumatoid factors (RF) [1], association with particular haplotypes of the HLA Class II gene [6, 7], occasional familial clustering with other autoimmune diseases [8, 9], infiltration of self-reactive CD4$^+$ T-cells in synovial inflammation (synovitis), and successful induction of arthritis in animals by immunization with self-constituents, such as type II collagen, in potent adjuvant [10]. There are also many reports that environmental agents, especially viruses, may directly affect the joint, causing chronic joint inflammation [11]. Fundamental questions in understanding the etiology of RA would therefore be how synovitis is initiated, in particular whether it is an autoimmunity mediated by self-reactive CD4$^+$ T-cells; if this is the case, how then arthritogenic self-reactive CD4$^+$ T-cells are generated and activated; and how genetic and environmental factors contribute to their generation and activation. Animal models of RA, in particular spontaneous models, are instrumental in addressing these questions. We have recently established a mouse strain that spontaneously develops CD4$^+$ T-cell-mediated chronic autoimmune arthritis immunopathologically resembling human RA [12]. The primary cause of the disease is a point mutation of the gene encoding ZAP-70, a key signal

The Hereditary Basis of Rheumatic Diseases, edited by Rikard Holmdahl transduction molecule in T-cells [13]. In this article, we discuss how this mono-genetic defect in T-cells, not in the joint, leads to the generation and activation of arthritogenic self-reactive CD4+ T-cells, what roles environmental factors play in the development of arthritis in the presence of this genetic abnormality, and how the findings in this model could be extended to our understanding of the etiology of human RA.

Immunopathology of arthritis in SKG mice

The SKG strain, which spontaneously develops chronic arthritis, is derived from our closed breeding colony of BALB/c mice. It shows the following clinical and immunopathological features [12]. Clinically, hyperemia becomes macroscopically evident around 2 months of age, initially at a few interpharangeal joints of the fore paws, then progressing in a symmetrical fashion to swelling of other finger joints of the fore and hind paws, and larger joints (wrists and ankles). Although swelling of small joints sometimes shows remission, swelling of large joints does not, eventual-ly resulting in ankylosis of wrists and ankles by 8–12 months of age due to destruc-tion and fusion of the subchondral bones, joint dislocation, and osteoporosis. Irre-spective of suffering from such severe chronic arthritis, the majority of SKG mice survive well to 1 year of age generally with more severe arthritides in females.

Histologically, SKG arthritis shows severe synovitis with massive subsynovial infiltration of neutrophils, lymphocytes, macrophages and plasma cells, villus pro-liferation of synoviocytes accompanying pannus formation and neovascularization, and neutrophil-rich exudates in the joint cavity. With progression of synoviocyte proliferation, pannus erodes adjacent cartilage and subchondral bone. CD4+ T-cells predominantly infiltrate the subsynovial tissue.

Serologically, SKG mice develop high titers of rheumatoid factor (RF), autoanti-bodies specific for type II collagen, antibodies reactive with heat shock protein (HSP)-70 of *Mycobacterium tuberculosis* presumably due to cross-reaction with a conserved epitope of HSP, severe hypergammaglobulinemia, and high concentration of circulating immune complexes, but no significant titer of anti-DNA antibodies or organ-specific autoantibodies such as anti-thyroglobulin autoantibody. Some SKG mice develop anti-cyclic citrullinated peptide (CCP) antibody (M. Hashimoto et al., unpublished data) [14].

As extra-articular manifestations of the disease, the majority (>90%) of mice older than 6 months of age develop interstitial pneumonitis with various degrees of perivascular and peribronchiolar cellular infiltration; more than 90% show infiltra-tion of inflammatory cells in the skin. Some (10–20%) mice had subcutaneous necrobiotic nodules, not unlike rheumatoid nodules in RA, and vasculitides. SKG mice do not develop lymphoadenopathy or lupus-like diseases (such as immune-complex glomerulonephritis).

Proinflammatory cytokines such as TNF-α, IL-1, IL-6 are abundantly produced in the affected joints [15]. Furthermore, the incidence and severity of SKG arthritis is significantly reduced when the mice are rendered TNF-α, IL-1 or IL-6-deficient, similar to the effects of anti-cytokine therapy in human RA [2, 15]. Notably, cytokine-deficient SKG mice free of arthritis still develop RF and anti-CCP antibody, indicating that these autoantibodies are not the consequence of joint inflammation ([15], M. Hashimoto, unpublished data) (see below).

Thus, spontaneous autoimmune disease in SKG mice resembles RA in clinical, histological, and serological characteristics of arthritis, the presence of extra-articular lesions, and the roles of proinflammatory cytokines in arthritis.

Thymic production of arthritogenic autoimmune T-cells in SKG mice

T-cells primarily mediate SKG arthritis [12]. Transfer of spleen and lymph node T-cells, CD4+ T-cells in particular, from arthritic SKG mice produces similar arthritis in T-cell-deficient athymic BALB/c nude mice, whereas transfer of the sera from the same SKG mice does not. Transfer of thymocyte suspensions from arthritic or non-arthritic young SKG mice also elicits severe arthritis in BALB/c nude mice and T/B cell-deficient C.B-17 SCID mice. In addition, transfer of T-cell-depleted bone marrow (BM)-cell suspensions from SKG mice produces severe arthritis in SCID mice, but not in SCID mice thymectomized prior to the transfer. Taken together, the RA-like arthritis in SKG mice is a *bona fide* autoimmune disease mediated by apparently joint-specific CD4+ T-cells, in accord with the infiltration of CD4+ T-cells to sub-synovial tissue in SKG arthritis. The SKG thymus is continuously generating arthritogenic autoimmune T-cells. Furthermore, SKG bone marrow cells give rise to such arthritogenic T-cells through the normal thymic environment, indicating that the arthritogenic abnormality in SKG mice is T-cell-intrinsic.

A point mutation of the ZAP-70 gene as the primary cause of SKG arthritis

The primary cause of SKG arthritis is a genetic abnormality inherited in an autosomal recessive fashion with nearly 100% penetrance of the trait in homozygotes raised in our conventional environment. Linkage analysis between the development of macroscopically evident arthritis and the homozygosity of chromosome-specific microsatellite markers maps the *skg* locus to the centromeric portion of chromosome 1 with the lod score of the locus as infinite. Construction of a high-resolution genetic map of the *skg* region, screening of yeast or bacterial artificial chromosome libraries for clones that cover the putative *skg* locus, and sequencing of such clones revealed a homozygous G-to-T substitution at nucleotide 489 in the SKG ZAP-70 gene, which alters codon163 from tryptophan to cysteine (W163C) (Fig. 1). The

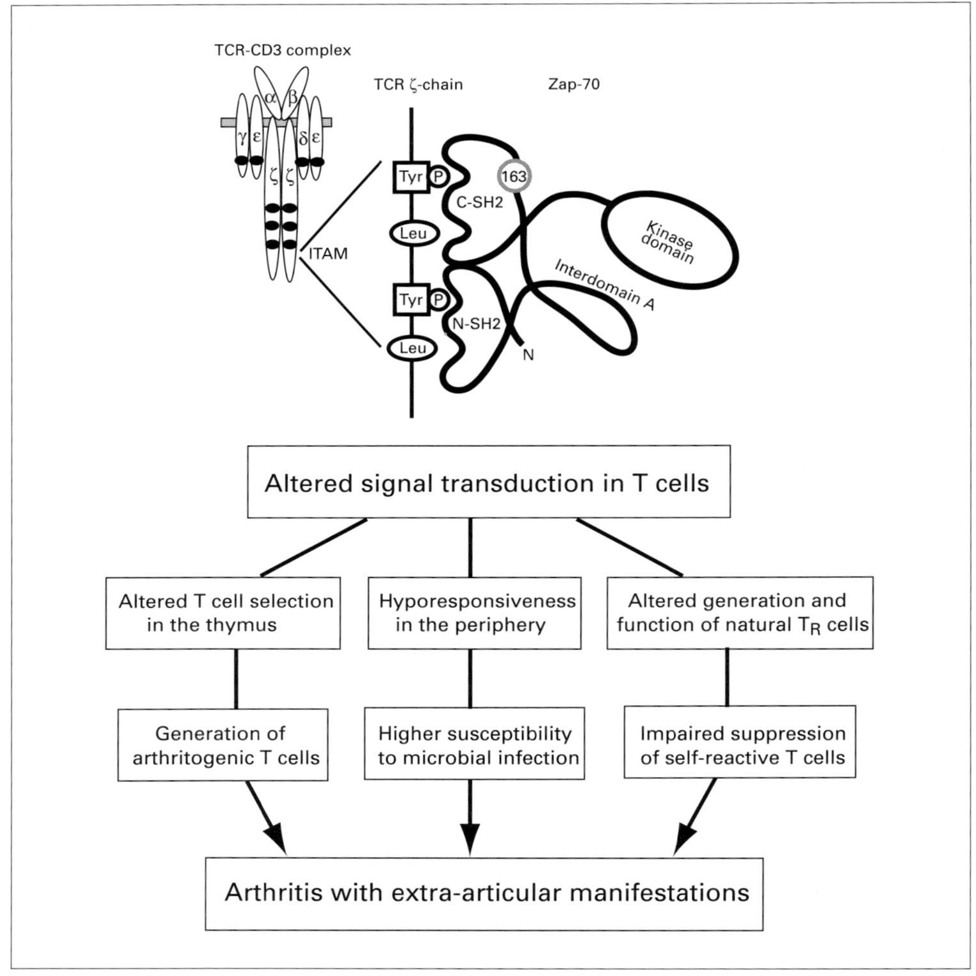

Figure 1

The ZAP-70 gene mutation and the mechanism of arthritis development in SKG mice.

position of the mutation corresponds to the initial amino acid residue of the C-terminal SH2 (SH2C) domain of ZAP-70 [16]. The revealed ZAP-70^{W163C} missense mutation is primarily responsible for SKG arthritis because ZAP-70-*skg/–* mice, made by mating ZAP-70-deficient (ZAP-70$^{–/–}$) mice with SKG (ZAP-70-*skg/skg*) mice, develop arthritis whereas ZAP-70-*skg/+* mice do not. In addition, T-cell-specific transgenic expression of the normal human ZAP-70 gene under the *lck* proximal promoter completely inhibits the development of arthritis in SKG mice whereas non-Tg littermates develop arthritis at 100% incidence. Thus, the ZAP-70^{W163C} alteration is not a mere polymorphism but the causative mutation of autoimmune arthritis in SKG mice.

As the consequence of the ZAP-70^{W163C} mutation, the tyrosine-phosphorylation status of major signal transduction molecules, including ZAP-70, TCR-ζ, LAT (linker for the activation of T-cells) and PLC-γ1, is extremely low in thymocytes and T-cells [17–19]. Calcium mobilization, one of the earliest events induced by TCR stimulation, is also greatly impaired in SKG thymocytes and T-cells presumably due to the reduced phosphorylation of PLC-γ1. Immunoprecipitation of ZAP-70 from stimulated SKG thymocytes and T-cells fail to co-precipitate tyrosine-phosphorylated p21 and p23 isoforms of TCR-ζ, indicating a defective interaction between ZAP-70 and TCR-ζ. SKG T-cells show no enhancement of the expression of Syk, another protein tyrosine kinase which might compensate the role of ZAP-70 in T-cells (T. Takahashi, unpublished data). Taken together, the ZAP-70^{W163C} mutation may impair the recruitment and association of ZAP-70 to tyrosine-phosphorylated immunoreceptor tyrosine-based activation motifs (ITAMs) of TCR-ζ and CD3 chains, thereby altering signal transduction through ZAP-70. Analysis of downstream MAP kinase activation in SKG thymocytes upon *in vitro* TCR stimulation shows that activation levels of ERK1/2 and p38 MAP kinases are less than 50% of control BALB/c thymocytes in every activation phase [20]. The kinetics of activation of both p54 and p46 JNK isoforms is also delayed in SKG thymocytes compared with BALB/c thymocytes. Thus, SKG thymocytes exhibit hypo-activation of three families of MAPKs upon TCR stimulation [12]. The attenuated ERK and P38/JNK pathways would impair positive and negative selection, respectively, of T-cells in SKG mice (see below) [21].

Altered T-cell development, function, and selection in SKG mice

The ZAP-70^{W163C} mutation and resultant altered signal transduction in T-cells then affects thymic development and differentiation of T-cells, in particular their positive and negative selection, responsiveness of mature T-cells to self and non-self antigens, and the generation and function of CD25$^+$CD4$^+$ natural regulatory T (T$_R$) cells [22] (Fig. 1).

Compared with normal BALB/c mice, TCR^{high} mature thymocytes decrease in number and ratio whereas TCR^{low} immature thymocytes increase in the SKG thymus, although the levels of TCR expression on mature thymocytes and peripheral T-cells are comparable between SKG and BALB/c. The number of peripheral CD4$^+$ and CD8$^+$ T-cells also decreases in SKG mice with a relative increase of B-cells.

Functionally, SKG T-cells are hyporesponsive to *in vitro* polyclonal T-cell stimulation, for example, with anti-CD3 monoclonal antibody (mAb), whereas they respond well to phorbol myristate acetate and ionomycin stimulation, indicating altered transduction of TCR-proximal signals in SKG T-cells. When the *skg* gene is homozygously introduced to DO11.10 TCR-transgenic (Tg) mice, in which the majority of T-cells express Tg TCRs specific for an ovalbumin (OVA) peptide, T-cells from such mice (designated DO.SKG mice) require ten times higher peptide concentrations to exhibit an equivalent magnitude of proliferation as DO11.10 T-cells. Furthermore, SKG thymocytes are less sensitive than BALB/c thymocytes to apoptosis induced by *in vivo* and *in vitro* TCR stimulation, in contrast to equivalent susceptibility to dexamethasone-induced apoptosis.

The ZAP-70^{W163C} mutation notably affects thymic positive and negative selection of T-cells. In DO.SKG (i.e., DO$^{skg/skg}$) mice, the number of CD4$^+$CD8$^-$ mature thymocytes decreases to ~20% of normal DO11.10 mice. Furthermore, only ~30% of CD4$^+$CD8$^-$ thymocytes and CD4$^+$ peripheral T-cells are Tg-TCR^{high} in contrast to ~80% in DO11.10 mice (Fig. 2a); the rest of CD4$^+$CD8$^-$ thymocytes/T-cells (i.e., Tg-TCR^{low} thymocytes/T-cells) in DO$^{skg/skg}$ mice appear to express endogenous TCR α-chains associated with transgenic β-chains, as the majority of their T-cells express transgenic β-chains at equivalent levels to DO11.10 mice. In *skg*-heterozygous DO$^{skg/+}$ mice, the reduction of the percentage of Tg-TCR^{high} T-cells and increase in Tg-TCR^{low} T-cells in the thymus and the periphery are intermediate between DO$^{skg/skg}$ and DO11.10 mice (Fig. 2a). Similar findings were also made with HY-TCR Tg mice that express transgenic TCRs specific for male-specific HY antigens on an H-2^b background (Fig. 2b). Female HY-TCR Tg mice positively select HY-TCR$^+$CD8$^+$CD4$^-$ thymocytes, whereas female HY-TCR Tg mice with the homozygous *skg* gene (designated HY-TCR.SKG Tg mice) on an H-2^b background hardly show positive selection of HY-TCR$^+$CD8$^+$CD4$^-$ thymocytes and T-cells. On the other hand, in contrast to substantial deletion of HY-TCR$^+$CD8$^+$CD4$^-$ thymocytes in male HY-TCR Tg mice, male HY-TCR.SKG Tg mice show efficient positive selection of HY-TCR$^+$CD8$^+$CD4$^-$ thymocytes. Notably, both DO.SKG and HY-TCR.SKG mice develop arthritogenic T-cells expressing endogenous TCR α-chains paired with transgenic β-chains, and consequently autoimmune arthritis at high incidences, due to positive selection of self-reactive T-cells that would not be positively selected in normal TCR Tg mice.

Figure 2
Altered thymic T-cell selection in SKG mice.
A. Staining of thymocytes or lymph node cells from a 4-month-old DO11.10, DO$^{skg/skg}$, or DO$^{skg/+}$ mouse with indicated mAbs. KJ1.26 is a mAb specific for the transgenic TCR. Dot-plot figures are scaled logarithmic. Ordinates of histograms denote cell number (arbitrary units). B. Staining of thymocytes or lymph node cells from a 4-month-old female or male HY-TCR or HY-TCR.SKG transgenic mouse with indicated mAbs. C. Selection shift of the T-cell repertoire and thymic production of arthritogenic T-cells.

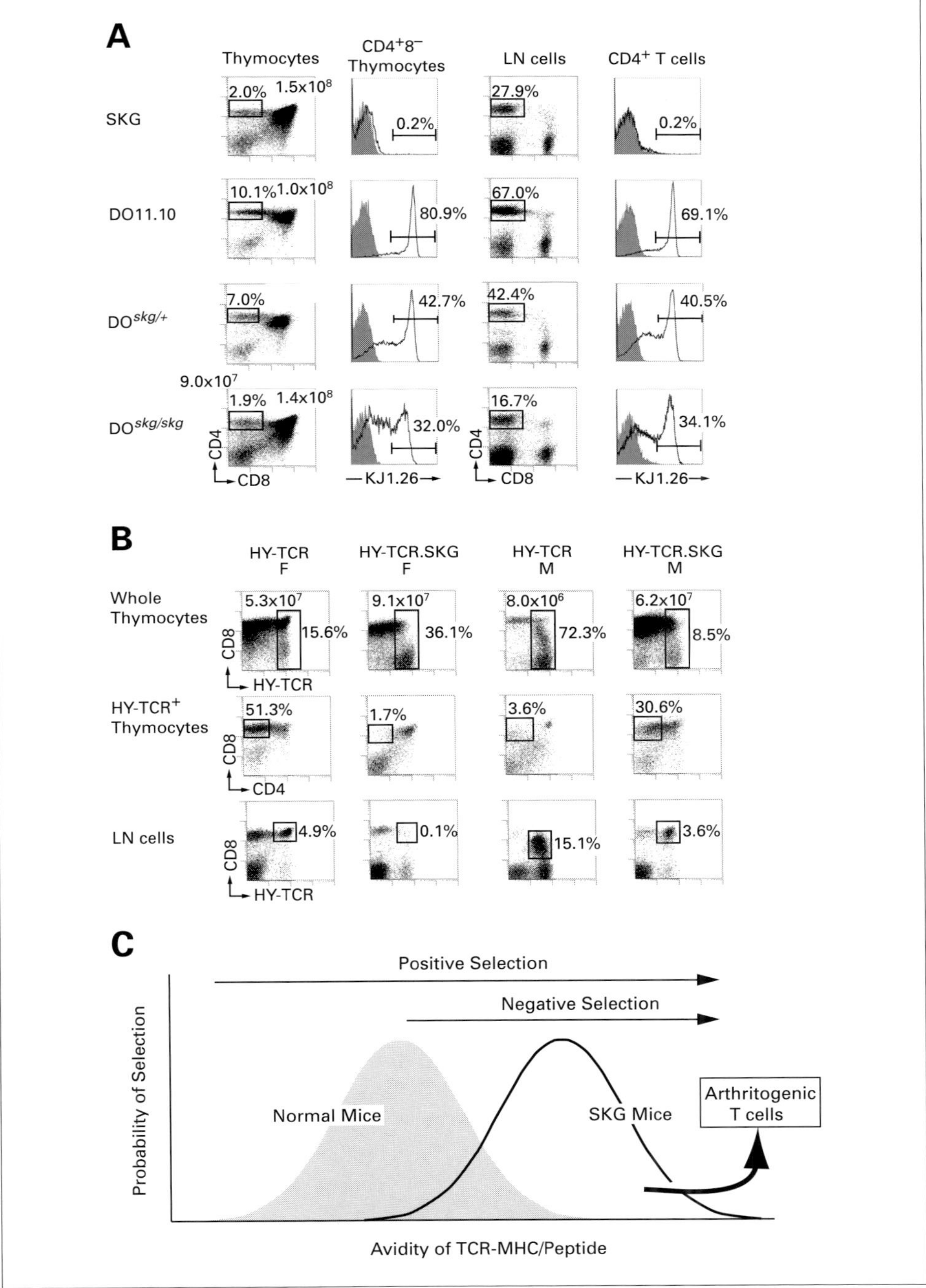
A
Thymocytes
CD4+8− Thymocytes
LN cells
CD4+ T cells
SKG
2.0% 1.5x10^8
0.2%
27.9%
0.2%
DO11.10
10.1% 1.0x10^8
80.9%
67.0%
69.1%
DO skg/+
7.0%
42.7%
42.4%
40.5%
9.0x10^7
DO skg/skg
1.9% 1.4x10^8
32.0%
16.7%
34.1%
CD4
CD8
KJ1.26
CD4
CD8
KJ1.26
B
HY-TCR F
HY-TCR.SKG F
HY-TCR M
HY-TCR.SKG M
Whole Thymocytes
5.3x10^7
15.6%
9.1x10^7
36.1%
8.0x10^6
72.3%
6.2x10^7
8.5%
CD8
HY-TCR
HY-TCR+ Thymocytes
51.3%
1.7%
3.6%
30.6%
CD8
CD4
LN cells
4.9%
0.1%
15.1%
3.6%
CD8
HY-TCR
C
Positive Selection
Negative Selection
Probability of Selection
Normal Mice
SKG Mice
Arthritogenic T cells
Avidity of TCR-MHC/Peptide

Thus, assuming that a certain level of signal strength is required for positive or negative selection of developing T-cells in the thymus [21], possible compensation for reduced signaling through the mutated ZAP-70 will raise the threshold of TCR avidity for self-peptide/MHC ligands required for each selection event, leading to a 'selection shift' of the T-cell repertoire towards higher self-reactivity than normal; that is, positive selection of highly self-reactive T-cells and failure in their negative selection, leading to thymic production of pathogenetic self-reactive T-cells, including arthritogenic T-cells, that would not be produced by the normal thymi of ZAP-70-intact animals (Fig. 2c). Importantly, this recessive mutation affects T-cell selection even in heterozygotes, although to a lesser degree than homozygotes, suggesting that the heterozygotes may also harbor a larger number of, or potentially more, pathogenic arthritogenic T-cells than normal.

In addition to positive and negative selection of self-reactive T-cells, another key function of the thymus in self-tolerance is the production of $CD25^+CD4^+$ natural regulatory T_R cells, which actively suppress the activation and expansion of self-reactive T-cells in the periphery [22]. They express the transcription factor Foxp3, which control their development and function [22]. In SKG mice, the number of $Foxp3^+CD25^+CD4^+$ T-cells and their *in vitro* suppressive function assessed by polyclonal TCR activation is not significantly different from normal BALB/c mice. However, when $CD25^+CD4^+$ T-cells from non-arthritic SKG or BALB/c mice are co-transferred with SKG $CD25^-CD4^+$ T-cells to BALB/c nude mice, SKG $CD25^+CD4^+$ T-cells are much less effective in inhibiting the development of arthritis (S. Tanaka et al., unpublished data). In addition, inoculation of $CD25^+CD4^+$ T-cells from normal BALB/c mice prevents arthritis development in SKG mice, as also shown in other arthritis models [23–25]. The altered autoimmune-inhibitory activity of SKG T_R cells could be due to their possible defects in activation or altered TCR repertoire in recognizing self-antigens due to the ZAP-70 gene mutation, since their thymic generation also needs a high-avidity interaction with self-peptide/MHC complexes expressed in the thymus [22].

Contribution of environmental factors to the development of SKG arthritis

The $ZAP-70^{W163C}$ mutation alone is not sufficient, however, for triggering arthritis in SKG mice. In the presence of the genetic abnormality of ZAP-70, environmental factors are needed for triggering arthritis, as illustrated by the following findings. First, in contrast with a high incidence of severe arthritis in our conventional environment, SKG mice raised in a microbially clean environment fail to develop arthritis [26]. When the latter are transferred after weaning to a conventional environment, they start to develop the disease at a high incidence [26]. Second, under the arthritis-resistant clean condition, a single injection of zymosan, a crude yeast cell wall extract, or glucose polymer β-1, 3-D-glucans (β-glucans), which are the main constituents of zymosan, can provoke chronically progressing severe arthritis in SKG mice [26]. Polyinosinic-polycytidylic acid (Poly [I:C]), a double-stranded RNA characteristic of some viruses, also shows a mild arthritogenic effect in SKG mice. Third, in the linkage analysis of the genes determining the susceptibility to arthritis in SKG mice, there is a significant linkage with the *H-2* locus on chromosome 17, besides the *skg* locus on chromosome 1 [12]. The mice that developed arthritis even in an arthritis-resistant microbially clean environment showed much more frequent homozygosity of the H-2^d haplotype than heterozygosity.

Taken together, in the presence of the SKG mutation, certain microbes, such as fungi, bacteria, and viruses, can activate arthritogenic T-cells through stimulating innate immunity, thereby evoking chronic autoimmune arthritis. Furthermore, the MHC gene polymorphism plays a significant role in determining the susceptibility to SKG arthritis and this genetic susceptibility exerts its effects depending on environmental conditions.

SKG autoimmune arthritis as a model of human RA

As discussed above, clinical and immunopathological characteristics of SKG arthritis resemble those of RA in humans, making the strain a suitable model for studying human RA. A cardinal feature of this model is that a genetic defect of T-cells, but not the joint, leads to the development of autoimmune arthritis. Another is that in the presence of this genetic abnormality of the T-cell compartment, environmental stimuli are required for triggering arthritis. The ZAP-70 genetic defect affects the thymic generation of arthritogenic CD4$^+$ T-cells and T$_R$ cell-mediated suppressive control of their activation in the periphery while environmental agents act to activate them via stimulating innate immunity (Fig. 3). Then, how is this mechanism of SKG autoimmune arthritis relevant to the pathogenesis of RA?

Like RA, SKG disease is a systemic disease including arthritis and extra-articular lesions. The development of RF and anti-CCP autoantibody is not the consequence of arthritis but an effect of the SKG T-cell anomaly (Fig. 3, see above). One may ask then why SKG mice predominantly develop autoimmune arthritis, but not other autoimmune diseases such as type 1 diabetes (T1D). This could be attributed at least in part to unique features of synovial cells as the target of SKG autoimmunity and for that matter of RA. For example, unlike pancreatic β-cells as the target of T1D, the synovium, which is composed of macrophage-like type A and fibroblast-like type B synoviocytes, is intrinsically capable of producing proinflammatory cytokines (such as IL-1, IL-6 and TNF-α) and chemical mediators (such as matrix metalloproteinases) destroying cartilage and bone [1, 2]. It is devoid of basement membrane and tight junctions, allowing easy invasion of proliferating synoviocytes to the surrounding tissue [1]. In addition, they are highly sensitive to various immunological stimuli including cytokines, biological substances, and presumably

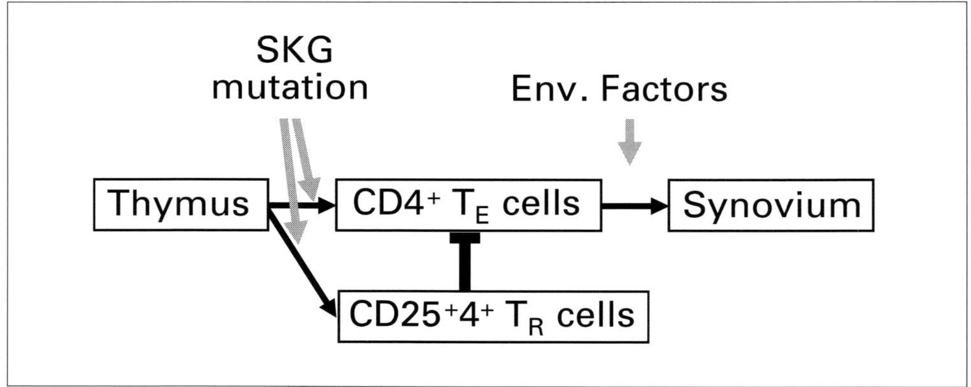

Figure 3
Contribution of genetic and environmental factors to the development of arthritis in SKG mice.

self-reactive T-cells, as typically illustrated by the predominant development of arthritis, but not other diseases, as a result of systemic overproduction of proinflammatory cytokines such as TNF-α or systemic deficiency of IL-1 receptor antagonist [27, 28]. It is therefore likely that a combination of high self-reactivity of SKG T-cells and high susceptibility of the synovium to inflammatory stimuli may lead to easy development of autoimmune arthritis in SKG mice irrespective of their formation of self-reactive T-cells with various other specificities. It remains to be determined whether T-cells responsible for causing interstitial pneumonitis or helping the formation of anti-CCP and other autoantibodies are also produced in SKG mice in parallel with the production of arthritogenic T-cells.

The ZAP-70^{W163C} mutation as the primary cause of SKG arthritis suggests that mutations of other loci of the ZAP-70 gene or the genes encoding other signaling molecules, especially at TCR proximal steps, may well contribute to the development of autoimmune arthritis by affecting a common signaling pathway(s). There is recent evidence that polymorphism of the PTPN22 gene, which encodes a hematopoietic-specific protein tyrosine phosphatase, also contributes to the risk of RA [29] and other autoimmune diseases including T1D, SLE, and Graves' disease [30–32]. The risk allele can alter the threshold of T-cell activation [30]. It remains to be determined whether the PTPN22 encoded by the risk allele affects thymic T-cell selection and/or the function of regulatory T-cells. It is also likely that variants of more then one gene encoding T-cell signaling molecules, including PTPN22, may have additive effects in altering T-cell signal transduction and thereby raising the genetic risk to RA.

It is well documented that MHC polymorphisms contribute to determining the genetic susceptibility to RA [6, 7]. SKG mice could be a good model for investigating the role of MHC in RA, especially in relation to the influence of the environment on arthritis development, as discussed above.

There is epidemiological evidence that environmental factors play significant roles in the development of RA [1, 7, 11]. Various environmental stimuli including viral infections can cause synovitis, which is usually self-limiting [11]. Administration of zymosan and β-glucan can indeed elicit in normal mice a transient arthritis, which is not T-cell-mediated, in contrast with zymosan- or β-glucan-triggered chronic T-cell-mediated autoimmune arthritis in SKG mice [26]. These findings when taken together indicate that synovial inflammation *per se* may not be sufficient to evoke chronic T-cell-mediated autoimmune arthritis unless one harbors arthritogenic T-cells sufficient in number and/or TCR specificity to mediate arthritis. The findings also suggest that one-time 'hit-and-run' exposure to an arthritogenic environmental agent capable of strongly stimulating innate immunity may suffice to trigger chronic arthritis in those who have already produced and harbor arthritogenic autoimmune T-cells due to genetic anomaly or variation.

Further genetic and immunological analyses of SKG arthritis at each step of the pathogenetic pathway from the ZAP-70 mutation, through thymic generation and peripheral activation of autoimmune T-cells, to inflammatory destruction of the joint, will help to elucidate how a specific combination of genetic and environmental factors leads to the development of RA. This will help then to design novel measures of detecting, treating and preventing RA.

Acknowledgements
The authors thank Dr. Zoltan Fehervari for critical reading of the manuscript. This work was supported by grants-in-aid from the Ministry of Education, Science, Sports and Culture, the Ministry of Human Welfare, and the Science and Technology Agency of Japan.

References

1   Harris ED (1997) *Rheumatoid arthritis*. W.B. Saunders, Philadelphia
2   Firestein GS (2003) Evolving concepts of rheumatoid arthritis. *Nature* 423: 356–336
3   Arend WP (2001) Physiology of cytokine pathways in rheumatoid arthritis. *Arthritis Care and Research* 45: 101–106
4   Feldmann M, Maini RN (2001) Anti-TNF alpha therapy of rheumatoid arthritis: what have we learned? *Annu Rev Immunol* 19: 163–196
5   Naka T, Nishimoto N, Kishimoto T (2002) The paradigm of IL-6: from basic science to medicine. *Arthritis Res* 4: S233–242
6   Buckner JH, Nepom GT (2002) Genetics of rheumatoid arthritis: is there a scientific

explanation for the human leukocyte antigen association? *Curr Opin Rheumatol* 14: 254–259

7   Gregersen PK (1999) Genetics of rheumatoid arthritis: confronting complexity. *Arthritis Res* 1: 37–44

8   Lin JP, Cash JM, Doyle SZ, Peden S, Kanik K, Amos CI, Bale SJ, Wilder RL (1998) Familial clustering of rheumatoid arthritis with other autoimmune diseases. *Hum Genet* 103: 475–482

9   Torfs CP, King MC, Huey B, Malmgren J, Grumet FC (1986) Genetic interrelationship between insulin-dependent diabetes mellitus, the autoimmune thyroid diseases, and rheumatoid arthritis. *Am J Hum Genet* 38: 170–187

10 Stuart JM, Townes AS, Kang AH (1984) Collagen autoimmune arthritis. *Annu Rev Immunol* 2: 199–218

11 Krause A, Kamradt T, Burmester GR (1996) Potential infectious agents in the induction of arthritides. *Curr Opin Rheumatol* 8: 203–209

12 Sakaguchi N, Takahashi T, Hata H, Nomura T, Tagami T, Yamazaki S, Sakihama T, Matsutani T, Negishi I, Nakatsuru S, Sakaguchi S (2003) Altered thymic T-cell selection due to a mutation of the ZAP-70 gene causes autoimmune arthritis in mice. *Nature* 426: 454–460

13 Chan AC, Iwashima M, Turck CW, Weiss A (1992) ZAP-70: a 70 kd protein-tyrosine kinase that associates with the TCR zeta chain. *Cell* 71: 649–662

14 Schellekens GA, de Jong BA, van den Hoogen FH, van de Putte LB, van Venrooij WJ (1998) Citrulline is an essential constituent of antigenic determinants recognized by rheumatoid arthritis-specific autoantibodies. *J Clin Invest* 101: 273–281

15 Hata H., Sakaguchi N, Yoshitomi H, Iwakura Y, Sekikawa K, Azuma Y, Kanai C, Moriizumi E, Nomura T, Nakamura T, Sakaguchi S (2004) Distinct contribution of IL-6, TNF-a, IL-1, and IL-10 to T cell-mediated spontaneous autoimmune arthritis in mice. *J Clin Invest* 114: 582–588

16 Hatada MH, Lu X, Laird ER, Green J, Morgenstern JP, Lou M, Marr CS, Phillips TB, Ram MK, Theriault K, et al (1995) Molecular basis for interaction of the protein tyrosine kinase ZAP-70 with the T-cell receptor. *Nature* 377: 32–38

17 Iwashima M, Irving BA, van Oers NS, Chan AC, Weiss A (1994) Sequential interactions of the TCR with two distinct cytoplasmic tyrosine kinases. *Science* 263: 1136–1139

18 Zhang W, Sloan-Lancaster J, Kitchen J, Trible RP, Samelson LE (1998) LAT: the ZAP-70 tyrosine kinase substrate that links T cell receptor to cellular activation. *Cell* 92: 83–92

19 van Oers NS, Tohlen B, Malissen B, Moomaw CR, Afendis S, Slaughter CA (2000) The 21- and 23-kD forms of TCR zeta are generated by specific ITAM phosphorylations. *Nature Immunol* 1: 322–328

20 Rincon M (2001) MAP-kinase signaling pathway in T cells. *Curr Opin Immunol* 13: 339–345

21 Starr TK, Jameson SC, Hogquist KA (2003) Positive and negative selection of T cells. *Annu Rev Immunol* 21: 139–176

22   Sakaguchi S (2004) Naturally arising CD4⁺ regulatory t cells for immunologic self-tolerance and negative control of immune responses. *Annu Rev Immunol* 22: 531–562

23   Morgan ME, Flierman R, van Duivenvoorde LM, Witteveen HJ, van Ewijk W, van Laar JM, de Vries RR, Toes RE (2005) Effective treatment of collagen-induced arthritis by adoptive transfer of CD25⁺ regulatory T cells. *Arthritis Rheum* 52: 2212–2221

24   Treschow AP, Backlund J, Holmdahl R, Issazadeh-Navikas S (2005) Intrinsic tolerance in autologous collagen-induced arthritis is generated by CD152-dependent CD4⁺ suppressor cells. *J Immunol* 174: 6742–6750

25   Londei M (2005) Role of regulatory T cells in experimental arthritis and implications for clinical use. *Arthritis Res Ther* 7: 118–120

26   Yoshitomi H, Sakaguchi N, Kobayashi K, Brown GD, Tagami T, Sakihama T, Hirota K, Tanaka S, Nomura T, Miki I et al (2005) A role for fungal β-glucans and their receptor Dectin-1 in the induction of autoimmune arthritis in genetically susceptible mice. *J Exp Med* 201: 949–960

27   Keffer J, Probert L, Cazlaris H, Georgopoulos S, Kaslaris E, Kioussis D, Kollias G (1991) Transgenic mice expressing human tumour necrosis factor: a predictive genetic model of arthritis. *EMBO J* 10: 4025–4031

28   Horai R, Saijo S, Tanioka H, Nakae S, Sudo K, Okahara A, Ikuse T, Asano M, Iwakura Y (2000) Development of chronic inflammatory arthropathy resembling rheumatoid arthritis in interleukin 1 receptor antagonist-deficient mice. *J Exp Med* 191: 313–320

29   Begovich AB, Carlton VE, Honigberg LA, Schrodi SJ, Chokkalingam AP, Alexander HC, Ardlie KG, Huang Q, Smith AM, Spoerke JM et al (2004) A missense single-nucleotide polymorphism in a gene encoding a protein tyrosine phosphatase (PTPN22) is associated with rheumatoid arthritis. *Am J Hum Genet* 75: 330–337

30   Bottini N, Musumeci L, Alonso A, Rahmouni S, Nika K, Rostamkhani M, MacMurray J, Meloni GF, Lucarelli P, Pellecchia M et al (2004) A functional variant of lymphoid tyrosine phosphatase is associated with type I diabetes. *Nat Genet* 36: 337–338

31   Kyogoku C, Langefeld CD, Ortmann WA, Lee A, Selby S, Carlton VE, Chang M, Ramos P, Baechler EC, Batliwalla FM et al (2004) Genetic association of the R620W polymorphism of protein tyrosine phosphatase PTPN22 with human SLE. *Am J Hum Genet* 75: 504–507

32   Velaga MR, Wilson V, Jennings CE, Owen CJ, Herington S, Donaldson PT, Ball SG, James RA, Quinton R, Perros P, Pearce SH (2004) The codon 620 tryptophan allele of the lymphoid tyrosine phosphatase (LYP) gene is a major determinant of Graves' disease. *J Clin Endocrinol Metab* 89: 5862–5865

A genetic approach to select and validate new targets for treatment of rheumatic diseases

Peter J. Olofsson

Arexis AB, Arvid Wallgrens Backe 20, 41346 Göteborg, Sweden

Introduction

Rheumatoid arthritis (RA) is a complex disease that despite decades of research still remains of unknown aetiology [1–3]. Key factors of the pathology are: the mechanism of induction of inflammation and the immunological network leading to disease perpetuation and development into a chronic severe disabling and painful disease. If answers to the fundamental complex molecular events that leads to the development of RA could be precisely dissected and analysed like an exact multi dimensional puzzle, this would enable academic scientists and pharmaceutical companies to identify new important targets for treatment of this disorder. To reach this goal, it is crucial to identify and quantify the gene expression of cells in the tissues that are involved in the initiation and development of chronic inflammatory arthritis. Although an approach focused on understanding the relevance of individual genes is important. There must also be efforts to understand the complete pathogenesis of RA. Hence, a real effort has to be made to avoid a too restrictive vision. Since articular diseases like RA are multifactorial disorders, a variety of unbiased possibilities of molecular mechanisms must be taken into account during identification of possible pharmaceutical drugs against RA.

In this chapter we will attempt to describe molecular genetic tools that can be used to identify novel genes that are involved in disease regulation and could be potential targets for drug development. The use of animal models for identification of new disease mechanism will be highlighted as a potential way to circumvent the heterogeneous complexity of the human population. We will also describe the potential to use the accumulating information of the human genome that can be used to individualise treatments against RA to increase treatment efficacy and avoid serious adverse effects of treatment in susceptible individuals.

The Hereditary Basis of Rheumatic Diseases, edited by Rikard Holmdahl
© 2006 Birkhäuser Verlag Basel/Switzerland

Pharmaceutical treatment of RA

The earliest treatments of rheumatoid arthritis are mainly directed against the symptoms of the disease and are generally aimed at maintaining the patient's quality of life. The drugs used are often non-steroidal anti-inflammatory drugs (NSAIDs), such as aspirin, naproxen and the cyclooxygenase 2 (COX2) inhibitors, all of which reduce pain and inflammation. NSAIDs are often used together with steroids (glucocortocoids), which together offer a very potent short-term anti-inflammatory effect.

Development of new treatments of RA has lead to the development of treatments that are directed more against the mechanism of the disease. These treatments not only aim at relieving the symptoms but also to stop the progression of the disease, and also if possible, revert the destruction of the involved cartilage joints. These treatments are collectively named disease modifying antirheumatic drugs (DMARDs). The most popular DMARD is methotrexate (MTX), a folic acid antagonist originally used for the treatment of cancer. Other DMARDs include sulphasalazine, tetracyclines and cyclosporine. Based on long-time use and positive outcome of treatments, MTX has generally been accepted as the leading DMARD [4, 5]. Although being an effective drug, about 50% of patients fail treatment due to side effects and loss of efficacy.

Hence, development of more potent drugs with similar mechanism as MTX, like leflonomide, has been approved for treatment of RA patients. However, there is still a large fraction of the RA patient group that do not get any symptom relief from these treatments.

Biologic drugs against RA

Due to the large number of patients that fail to have long lasting positive outcomes after treatments with drugs like MTX, leflunomide and corticosteroids new promising, but expensive, cytokine-targeted therapies have been developed. These treatments have been developed based on findings of the cytokine network that is involved in the immunological regulation of RA.

Most of these evolving therapies target cytokines of central importance in inflammatory conditions and include TNF-α trapping drugs, such as etanercept (a fusion protein combining the ligand binding portion of human TNF receptor 2 (p75) with the Fc-portion of human IgG) [6] and infliximab (a recombinant chimeric antibody against human TNF-α, consisting of human IgG1 constant and mouse variable regions) [7], as well as IL-1Rα (a recombinant human interleukin-1 receptor antagonist) [8].

In addition, treatment with these cytokine-trapping drugs can be improved by combination with DMARDs, as has been reported for etanercept combined with MTX [9].

The genetic component in RA

RA is known to depend both on environmental factors [10] and familial inheritance [11–13]. Estimations of both the genetic and the environmental impact have been extensively studied by measuring the risk for siblings divided by the risk for the population to develop disease. For example, in RA, the relative risk value varies between 2 and 17 [14], depending upon discrepancies for disease diagnosis as well as the incidence for RA that varies among different populations [15]. Another way to estimate the genetic heritability and the environmental impact of complex diseases is by comparing monozygotic twins (genetically identical) with dizygotic twins (approximately 50% genetically identical) and compare with the incidence in the general population. The heritability in RA has then been estimated to 60% [12]. However, a tendency of overestimation has been suggested for these twin studies, because of the shared environment. Genetic analysis of the inheritance of RA and other multifactorial diseases has for many years been complicated by the fact that only fractions of the genome sequence have been known. However, with the genetic revolution that have culminated during the last few years through the release of the full genome sequence of human [16, 17], great expectations are now set upon researchers to resolve complex inherited diseases like RA [14]. The formation of public genome databases, which today contain complete sequence of the human genome as well as the fast accumulation of genetic markers, will facilitate genetic association studies of complex diseases like RA [18].

Gene expression profiling to identify drug targets

New clinical programs for treatments of rheumatoid arthritis are more and more directed on single molecular targets in the complex immune system. The choice of these drug targets have evolved from a growing knowledge of the underlying mechanism of inflammatory diseases, where most of these targets focus on candidate genes in cytokine clusters [19]. Other examples of genes that have been implicated as potential targets for treatment or disease prognosis markers are HLA and MMP-3 [20].

However, the drawback of these targets is that they all are dependent on previous knowledge of the studied molecular networks. As such, there is a deficit in identification of novel targets from previously unknown or overlooked perspectives on disease induction and perpetuation mechanisms involved in RA.

Ways to identify such a target lies in an unbiased analysis of a genetic molecular approach that precisely examines, unravels and dissects various aspects of the disease.

The first approach of choice to get a molecular insight in a complex biological system is a complete gene expression analysis including all annotated as well as yet un-annotated genes in the genome. With this approach one will detect important genes that are differentially regulated in aspect to the investigated condition for a specific tissue and time-point in the disease development. One strong aspect of this method is the large set of genes that can be analysed in the same experimental set-up (on the same chip) making it possible to use the produced clustering and pathway analysis of large sets of genes in, and make use of all publicly available information of, these genes in automated analyses [21, 22].

To reach a complete understanding of the molecular regulation of arthritis one must address the complete gene expression simultaneously. To reach this goal, it is crucial to identify and quantify the genes expressed by the cells that are present and interacting in the tissue.

However, one big disadvantage with this method is that the genes of future interest must be differentially expressed at a transcriptional level to be detected by this method. Another difficulty with this method lies in the problematic choice of time-point and tissue/cell type of interest for analysis. These kinds of experimental design are still very complex and time consuming and generate huge amounts of data. Hence, it is necessary before the experimental initiation to have a clear image of the time-point and tissue of interest for the study. A further difficulty is the need of biological replicates and the genetic heterogeneity that is involved in studies in patient material. Therefore, scepticisms to the use of gene expression profiling as a method to achieve an understanding of rheumatic diseases has been put forward [23]. However, recent analysis of gene expression fingerprints of individual patients with early rheumatoid arthritis might give hope for this still technically evolving diagnosis method [24, 25].

Linkage analysis to identify drug targets

The second approach of choice, to identify arthritis regulating genes for drug development, is population genetic studies by linkage association and positional cloning. With this method, genes of utmost importance for the disease development will be identified regardless of the level of expression, tissue specificity or time of expression. With positional cloning, totally new pathways for disease regulation can be discovered and genes previously not annotated or barely studied in another context can be discovered to be of relevance in the disease.

One drawback of these studies lies in the amount of individuals that have to be included in the investigation and the time it takes to analyse the linkage before positional cloning of individual genes can take place. For example it has been estimated that between 1,000–2,000 sibling pairs will be necessary to identify susceptibility genes with moderate effect [26, 27]. Due to these difficulties, few single genes have been identified from the large number of QTL (Quantitative Trait Loci) that have been identified to control RA. Positional cloning in humans is also hampered by genetic heterogeneity and lack of more individuals that are needed to confirm linkages and make the final localisation of individual disease causative genes.

Besides HLA, which has been identified in association studies and been estimated to account for one third of the inherited susceptibility of RA [28, 29], no other loci of significant contribution to RA has been detected despite several linkage analyses performed in human patient studies [30–33]. Multifactorial environmental influence, genetic and phenotypic heterogeneity together with the impracticality of obtaining individual samples in high enough number makes the majority of human linkage analyses of complex diseases predestined to end short of significant disease-associated chromosomal loci [34, 35]. Another major cumbersome problem that human geneticists have to face when dissecting complex traits is the impracticality to proceed further once a significant loci has been identified. The need for additional informative individuals might be an overwhelming obstacle as it is necessary to obtain additional samples within the same ethnic group as the original study [36]. Despite these difficulties there have been some occasions of successful identifications of autoimmunity regulating genes in humans [37–39], but overall it is difficult to identify genes in complex human diseases [40]. Furthermore, if a significant genetic association is identified, it is needed to identify candidate genes for positional cloning. Positional cloning approaches will most often demand additional large numbers of patient family members to identify single predisposing genes.

Genetically segregating crosses and positional cloning in animal models

One way of overcoming many of the problems that are mentioned in correlation with linkage studies in human populations could be solved by the use of animal models for identification of arthritis regulating genes.

For studies of arthritis, there exist several animal models. The most common model of arthritis is induced both in mice [41] and in rats [42] with collagen type II emulsified in incomplete Freund's adjuvant (i.e., mineral oil) or complete Freund's adjuvant (i.e., mycobacteria cell walls in mineral oil). Besides the arthritis induced with cartilage specific proteins like the type II collagen induced arthritis (CIIA) [41, 42], type XI collagen induced arthritis (CIXIA) [43] or the cartilage oligomeric matrix protein induced arthritis (COMPIA) [44], it is also possible to induce arthritis with mycobacterium emulsified in adjuvant (MIA), or for that matter even with mineral oil only (oil induced arthritis, OIA) or with synthetic adjuvants like pristane (pristane induced arthritis, PIA) [45–50]. All models of arthritis in mice and rats have different characteristics and upon comparison with criteria used for diagnosing RA [51] it is clear that the consensus knowledge of all these models can be used to study most aspects of RA.

There are several advantages in using animal models to study the genetic contribution in complex diseases. Firstly, many animals can be used in a single study where the environment is controlled. Secondly, inbred strains can be crossed using defined crossing procedures to provide controlled genetic segregation in the offspring in order to identify chromosomal regions, or even genes which regulate the disease. Thirdly, the function of identified chromosomal regions or isolated genes can be extensively studied by several techniques, for example transfer of bone marrow or inflammatory cells, as well as various treatments both preventive and therapeutic. Finally also, once a new drug target has been identified in one inbred animal strain, a direct proof of concept experiment can be designed in these animals to target the pathway by new treatment regimes. Such experiments performed in these well characterised and recognised arthritis models will strengthen further drug development into clinical trials.

Inbred rat and mouse strains have normally been inbred for at least 20 generations, which means that all loci are homozygous within each species, and there are several inbred strains that are either susceptible or resistant to a specific form of arthritis.

After the identification of significant QTL that regulates a specific arthritic condition in the animals it is easy, but time-consuming, to isolate the genomic region of interest in a congenic strain [52]. Established congenic strains in comparison with the parental strains could be used for 1) fine mapping of a genetic region by the creation of further recombinations; 2) comparative sequencing to identify polymorphisms; 3) expression analysis of selected proteins or mRNA; 4) functional studies of identified candidate genes.

Then again, when analysing genetic regulation of arthritis in animal models one must keep in mind the evolutionary difference between man and rodents, as well as the fact that an animal model of RA is not the same as the human disease. Hence, the most important finding, obtained from positional cloning of arthritis regulating genes in animal models, for future studies of RA is not the identified genes *per se*, but rather the identification of previously unknown pathological mechanisms and pathways of disease regulation. Since the first publication, in 1996, of genetic linkage analysis of CIA in rats [53] and mice [54, 55], studies of linkage analysis of arthritis in rats has been ongoing using various models of arthritis and different inbred mouse and rat strains (reviewed in [56–58]).

New targets identified by positional cloning in animal models

Genetic projects based on linkage analysis of complex disorders are time and lab consuming, especially as isolation of associated loci in congenic strains demands years of breeding. Hence, there have been doubts concerning the potential of reverse genetics and positional cloning of polygenic disease genes, both in humans and in animal models [26, 40]. Quite numerous numbers of QTL have been assigned in mouse and rat models of arthritis [56–58]. Some of these are now getting closer to identification of the arthritis regulating genes [59, 60]. The first of these genes to be positionally identified was the successful isolation of a functional polymorphism of Ncf1, explaining the effect on arthritis severity by the quantitative trait locus *Pia4*, through positional cloning in rats, and later also verified to be of importance in collagen induced arthritis in Ncf1 deficient mice [61, 62]. Positional identification of *Ncf1* was probably the first in a long range of genes that will be identified in animal models of arthritis, that all will be important pieces in a puzzle of the complex inheritance of arthritis. Eventual assembly of this puzzle will give deeper understanding of the pathogenesis of many autoimmune diseases. Identification of arthritis regulating genes, like *Ncf1*, in rodents can directly be transferred to analysis in human populations. In most cases genes in rodents will play a similar role in human arthritis. More important, however, is to use this information to address the molecular pathways surrounding the identified gene in RA. In the case of *Ncf1* it was found that the capacity to produce radical oxygen species was of utmost importance for the regulation of autoreactive T-cells in rats. Interestingly, the susceptibility to adjuvant-induced arthritis was correlated to a reduced capacity of the NADPH oxidase to produce radical oxygen species. With this approach new mechanisms for regulation of arthritis are revealed. By investigating the relevance of the new pathways in the RA and other autoimmune conditions, as well as in other animal models, novel targets for drug development will be displayed.

Pharmacogenomics to tailor RA treatment

RA is a disease syndrome with so far unknown aetiology. The large heterogeneity of the human population at large, as well as the heterogeneity of the RA patient population makes such diseases complex to diagnose genetically. The genetic influence of RA is quite modest, based on the concordance rate it has been suggested to be 12–15% [11, 13, 63]. However, the heritability of RA quantified from combined studies has been calculated to be as high as 60% [12].

The fact that there is a genetic preponderance to develop RA does however not necessarily mean that the disease will develop. This is known as *incomplete penetrance*. The same inheritable disease can be caused by more than one combination of genetic defects. This is known as *genetic heterogeneity*. Many genes influence complex disorders as well as normal physiological conditions. The presence of several predisposing genes may be necessary to develop a disease. This is known as *polygenic inheritance*. All these aspects of heritability are of importance in RA making RA a genetically complex disorder [64]. The genetic complexity of RA does not only complicate a correct diagnosis of the disease. It is also relevant for the treatment of RA patients. It is clear from clinical trials and outcome of treatment of RA patients that the efficacy of various treatments is highly individual.

Today methotrexate (MTX) is the most commonly used disease-modifying antirheumatic drug against RA [65]. Still, more than 50% of patients treated with only methotrexate or leflunomide fail to respond, or get additional treatment with a second DMARD within 18 months [66]. Not only is a failure to respond to MTX treatment a reason for limitation of using MTX: 10–30% of patients show toxic effects from MTX within 1 year of initiating MTX treatment. This represents a large fraction of patients for which MTX treatment may be unsuitable [67]. Despite the historical use of MTX for RA treatment there are today no reliable tests or assays that can predict the efficacy or toxicity of MTX. Also, with the treatment of recently-developed biological therapies like etanercept, infliximab and anakinra, a large proportion of the patients fail to respond well to the treatment. This is due to the fact that, for all available treatments for RA, there is always a patient group that fails to respond due to low efficacy or toxicity. In these instances, one has to consider the development of personalised or individual drug therapy.

The field of pharmacogenetics focuses on genetic polymorphisms to get a genetic understanding about drug metabolism and efficacy [68]. Pharmacogenetics is leading to new approaches to drug discovery, individualised drug therapy, and new perspectives on disease prevention. Traditional drug therapy treats all patients with the same disease as one large group, indifferent to individual differences to drug response. Pharmacogenetics should help focus therapies on smaller patient populations, which exhibit the same disease phenotype but are characterised by different genetic profiles. In the era of growing genetic information of the human genome together with accessible collections of patient samples with matched clinical diagnosis, there is a growing rational for investigations that address the possibility to individualise pharmacological treatment. For the development of future drugs that have to enter clinical trials to investigate efficacy and safety, this has tremendous impact on the outcome. Instead of the usual standard trials consisting of randomised matched groups one could envision that clinicians could stratify treatment groups with respect to their genetic profile. This approach will, of course, require large efforts in genome analysis. One example of such an approach is the geneologic characterisation of the Icelandic population that is performed by deCODE Genetics (Reykjavik, Iceland) [69]. Genetic analyses of complex diseases, like RA, are most efficiently addressed in genetically isolated populations where the heterogeneity is less pronounced [70, 71]. Likewise, design and performance of clinical trials could be performed in genetically characterised populations. Such clinical trials would then enable stratification of the material after efficacy and safety of the studied treatment. Such an analysis could give information about populations where the treatment is more likely to be effective as well as which patients are more likely to suffer from low tolerance. With this information to hand, extended clinical trials and market potential for the new drug would be more appreciated. Such approaches are also of importance in regard to the potential that some drugs might display different efficacy in different ethnical populations [72]. For some drugs, these

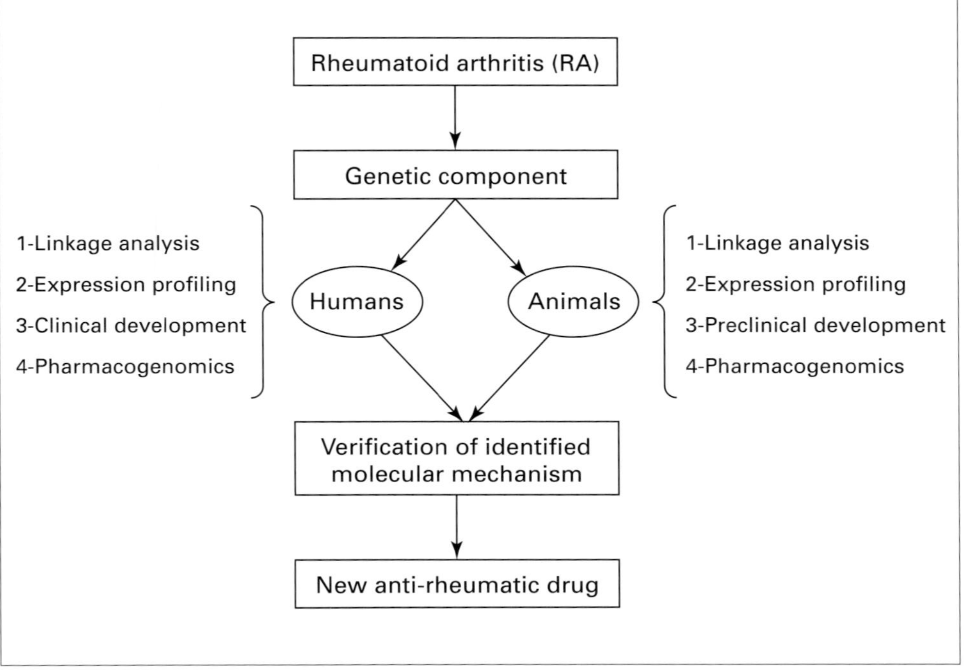

Figure 1
Schematic pathway showing the necessary interactions of studies of patients and animal models that is needed for identification of new drug targets, based on inheritance of RA, and the following development of new treatments against RA.

screening possibilities would increase the percentage of patients that tolerates the drug under prolonged time as well as having a larger extent of relieved symptoms, which will strengthen the market potential of the drug. Traditionally large pharmaceutical companies have not been motivated to pursue a pharmacogenetic approach because it would segment the market and weaken sales for their blockbuster drugs. However, with the higher attention paid to patients with serious side effects those blockbusters might be endangered, pharmacogenetics might be a way to save these drugs from market withdrawal. This would also be of utmost importance for drugs that are reported to have serious advert effects in minor patient groups. One way of keeping these drugs on the market would thus be a verified method to identify those patients that potentially would suffer from adverse effects, possibly by nature of a genetic difference, and to exclude them from the particular treatment. One would therefore envision the establishment in the future of the production of gene chips that may be used to individualise and optimise drug treatment.

References

1   Feldmann M, Brennan FM, Maini RN (1996) Rheumatoid arthritis. *Cell* 85: 307–310
2   Firestein GS (2003) Evolving concepts of rheumatoid arthritis. *Nature* 423: 356–361
3   Lee DM, Weinblatt ME (2001) Rheumatoid arthritis. *Lancet* 358: 903–911
4   Bannwarth B, Labat L, Moride Y, Schaeverbeke T (1994) Methotrexate in rheumatoid arthritis. An update. *Drugs* 47: 25–50
5   Pincus T, Marcum SB, Callahan LF (1992) Long-term drug therapy for rheumatoid arthritis in seven rheumatology private practices: II. Second line drugs and prednisone. *J Rheumatol* 19: 1885–1894
6   Franklin CM (1999) Clinical experience with soluble TNF p75 receptor in rheumatoid arthritis. *Semin Arthritis Rheum* 29: 172–181
7   Antoni C, Kalden JR (1999) Combination therapy of the chimeric monoclonal anti-tumor necrosis factor alpha antibody (infliximab) with methotrexate in patients with rheumatoid arthritis. *Clin Exp Rheumatol* 17: S73–77
8   Watt I, Cobby M (2001) Treatment of rheumatoid arthritis patients with interleukin-1 receptor antagonist: radiologic assessment. *Semin Arthritis Rheum* 30: 21–25
9   Klareskog L, van der Heijde D, de Jager JP, Gough A, Kalden J, Malaise M, Martin Mola E, Pavelka K, Sany J, Settas L et al (2004) Therapeutic effect of the combination of etanercept and methotrexate compared with each treatment alone in patients with rheumatoid arthritis: double-blind randomised controlled trial. *Lancet* 363: 675–681
10   Symmons DP, Bankhead CR, Harrison BJ, Brennan P, Barrett EM, Scott DG, Silman AJ (1997) Blood transfusion, smoking, and obesity as risk factors for the development of rheumatoid arthritis: results from a primary carebased incident case-control study in Norfolk, England. *Arthritis Rheum* 40: 1955–1961
11   Aho K, Koskenvuo M, Tuominen J, Kaprio J (1986) Occurrence of rheumatoid arthritis in a nationwide series of twins. *J Rheumatol* 13: 899–902
12   MacGregor AJ, Snieder H, Rigby AS, Koskenvuo M, Kaprio J, Aho K, Silman AJ (2000) Characterizing the quantitative genetic contribution to rheumatoid arthritis using data from twins. *Arthritis Rheum* 43: 30–37
13   Silman AJ, MacGregor AJ, Thomson W, Holligan S, Carthy D, Farhan A, Ollier WE (1993) Twin concordance rates for rheumatoid arthritis: results from a nationwide study. *Br J Rheumatol* 32: 903–907
14   Seldin MF, Amos CI, Ward R, Gregersen PK (1999) The genetics revolution and the assault on rheumatoid arthritis. *Arthritis Rheum* 42: 1071–1079
15   Wiles N, Symmons DP, Harrison B, Barrett E, Barrett JH, Scott DG, Silman AJ (1999) Estimating the incidence of rheumatoid arthritis: trying to hit a moving target? *Arthritis Rheum* 42: 1339–1346
16   Lander ES, Linton LM, Birren B, Nusbaum C, Zody MC, Baldwin J, Devon K, Dewar K, Doyle M, FitzHugh W et al (2001) Initial sequencing and analysis of the human genome. *Nature* 409: 860–921
17   Venter JC, Adams MD, Myers EW, Li PW, Mural RJ, Sutton GG, Smith HO, Yandell

M, Evans CA, Holt RA et al (2001) The sequence of the human genome. *Science* 291: 1304–1351

18   Foster MW, Sharp RR (2004) Beyond race: towards a whole-genome perspective on human populations and genetic variation. *Nat Rev Genet* 5: 790–796

19   John S, Eyre S, Myerscough A, Barrett J, Silman A, Ollier W, Worthington J (2001) Linkage and association analysis of candidate genes in rheumatoid arthritis. *J Rheumatol* 28: 1752–1755

20   Scott DL (2003) Genotypes and phenotypes: should genetic markers and clinical predictors drive initial treatment decisions in rheumatic diseases? *Curr Opin Rheumatol* 15: 213–218

21   Devauchelle V, Marion S, Cagnard N, Mistou S, Falgarone G, Breban M, Letourneur F, Pitaval A, Alibert O, Lucchesi C et al (2004) DNA microarray allows molecular profiling of rheumatoid arthritis and identification of pathophysiological targets *Genes Immun* 5: 597–608

22   Heller RA, Schena M, Chai A, Shalon D, Bedilion T, Gilmore J, Woolley DE, Davis RW (1997) Discovery and analysis of inflammatory disease-related genes using cDNA microarrays. *Proc Natl Acad Sci USA* 94: 2150–2155

23   Lanchbury J, Hall M, Steer S (2002) Progress and problems in defining susceptibility genes for rheumatic diseases. *Rheumatology (Oxford)* 41: 361–364

24   Jarvis JN, Centola M (2005) Gene-expression profiling: time for clinical application? *Lancet* 365: 199–200

25   Olsen N, Sokka T, Seehorn CL, Kraft B, Maas K, Moore J, Aune TM (2004) A gene expression signature for recent onset rheumatoid arthritis in peripheral blood mononuclear cells. *Ann Rheum Dis* 63: 1387–1392

26   Risch NJ (2000) Searching for genetic determinants in the new millennium. *Nature* 405: 847–856

27   Altshuler D, Hirschhorn JN, Klannemark M, Lindgren CM, Vohl MC, Nemesh J, Lane CR, Schaffner SF, Bolk S, Brewer C et al (2000) The common PPARgamma Pro12Ala polymorphism is associated with decreased risk of type 2 diabetes. *Nat Genet* 26: 76–80

28   Deighton CM, Walker DJ, Griffiths ID, Roberts DF (1989) The contribution of HLA to rheumatoid arthritis. *Clin Genet* 36: 178–182

29   Weyand CM, Goronzy JJ (2000) Association of MHC and rheumatoid arthritis. HLA polymorphisms in phenotypic variants of rheumatoid arthritis. *Arthritis Res* 2: 212–216

30   Cornelis F, Faure S, Martinez M, Prud'homme JF, Fritz P, Dib C, Alves H, Barrera P, de Vries N, Balsa A et al (1998) New susceptibility locus for rheumatoid arthritis suggested by a genome-wide linkage study. *Proc Natl Acad Sci USA* 95: 10746–10750

31   Hardwick LJ, Walsh S, Butcher S, Nicod A, Shatford J, Bell J, Lathrop M, Wordsworth BP (1997) Genetic mapping of susceptibility loci in the genes involved in rheumatoid arthritis. *J Rheumatol* 24: 197–198

32   Jawaheer D, Seldin MF, Amos CI, Chen WV, Shigeta R, Monteiro J, Kern M, Criswell LA, Albani S, Nelson JL et al (2001) A genomewide screen in multiplex rheumatoid

arthritis families suggests genetic overlap with other autoimmune diseases. *Am J Hum Genet* 68: 927–936

33 MacKay K, Eyre S, Myerscough A, Milicic A, Barton A, Laval S, Barrett J, Lee D, White S, John S et al (2002) Whole-genome linkage analysis of rheumatoid arthritis susceptibility loci in 252 affected sibling pairs in the United Kingdom. *Arthritis Rheum* 46: 632–639

34 Altmuller J, Palmer LJ, Fischer G, Scherb H, Wjst M (2001) Genomewide scans of complex human diseases: true linkage is hard to find. *Am J Hum Genet* 69: 936–950

35 Dahlman I, Eaves IA, Kosoy R, Morrison VA, Heward J, Gough SC, Allahabadia A, Franklyn JA, Tuomilehto J, Tuomilehto-Wolf E et al (2002) Parameters for reliable results in genetic association studies in common disease. *Nat Genet* 30: 149–150

36 Morahan G, Morel L (2002) Genetics of autoimmune diseases in humans and in animal models. *Curr Opin Immunol* 14: 803–811

37 Horikawa Y, Oda N, Cox NJ, Li X, Orho-Melander M, Hara M, Hinokio Y, Lindner TH, Mashima H, Schwarz PE et al (2000) Genetic variation in the gene encoding calpain-10 is associated with type 2 diabetes mellitus. *Nat Genet* 26: 163–175

38 Lesage S, Zouali H, Cezard JP, Colombel JF, Belaiche J, Almer S, Tysk C, O'Morain C, Gassull M, Binder V et al (2002) CARD15/NOD2 mutational analysis and genotype-phenotype correlation in 612 patients with inflammatory bowel disease. *Am J Hum Genet* 70: 845–857

39 Prokunina L, Castillejo-Lopez C, Oberg F, Gunnarsson I, Berg L, Magnusson V, Brookes AJ, Tentler D, Kristjansdottir H, Grondal G et al (2002) A regulatory polymorphism in PDCD1 is associated with susceptibility to systemic lupus erythematosus in humans. *Nat Genet* 28: 28

40 Rannala B (2001) Finding genes influencing susceptibility to complex diseases in the post-genome era. *Am J Pharmacogenomics* 1: 203–221

41 Courtenay JS, Dallman MJ, Dayan AD, Martin A, Mosedale B (1980) Immunisation against heterologous type II collagen induces arthritis in mice. *Nature* 283: 666–668

42 Trentham DE, Townes AS, Kang AH (1977) Autoimmunity to type II collagen an experimental model of arthritis. *J Exp Med* 146: 857–868

43 Lu S, Carlsen S, Hansson AS, Holmdahl R (2002) Immunization of rats with homologous type XI collagen leads to chronic and relapsing arthritis with different genetics and joint pathology than arthritis induced with homologous type II collagen. *J Autoimmun* 18: 199–211

44 Carlsen S, Hansson AS, Olsson H, Heinegard D, Holmdahl R (1998) Cartilage oligomeric matrix protein (COMP)-induced arthritis in rats. *Clin Exp Immunol* 114: 477–484

45 Holmdahl R, Goldschmidt TJ, Kleinau S, Kvick C, Jonsson R (1992) Arthritis induced in rats with adjuvant oil is a genetically restricted, alpha beta T-cell dependent autoimmune disease. *Immunology* 76: 197–202

46 Holmdahl R, Lorentzen JC, Lu S, Olofsson P, Wester L, Holmberg J, Pettersson U

(2001) Arthritis induced in rats with nonimmunogenic adjuvants as models for rheumatoid arthritis. *Immunol Rev* 184: 184–202

47   Lorentzen JC (1999) Identification of arthritogenic adjuvants of self and foreign origin. *Scand J Immunol* 49: 45–50

48   Pearson CM (1956) Development of arthritis, periarthritis and perioscitis in rats given adjuvants. *Proc Soc Exp Biol Med* 91: 91–101

49   Vingsbo C, Jonsson R, Holmdahl R (1995) Avridine-induced arthritis in rats; a T cell-dependent chronic disease influenced both by MHC genes and by non-MHC genes. *Clin Exp Immunol* 99: 359–363

50   Vingsbo C, Sahlstrand P, Brun JG, Jonsson R, Saxne T, Holmdahl R (1996) Pristane-induced arthritis in rats: a new model for rheumatoid arthritis with a chronic disease course influenced by both major histocompatibility complex and non-major histocompatibility complex genes. *Am J Pathol* 149: 1675–1683

51   Arnett FC, Edworthy SM, Bloch DA, McShane DJ, Fries JF, Cooper NS, Healey LA, Kaplan SR, Liang MH, Luthra HS et al (1988) The American Rheumatism Association 1987 revised criteria for the classification of rheumatoid arthritis. *Arthritis Rheum* 31: 315–324

52   Wakeland E, Morel L, Achey K, Yui M, Longmate J (1997) Speed congenics: a classic technique in the fast lane (relatively speaking). *Immunol Today* 18: 472–477

53   Remmers EF, Longman RE, Du Y, O'Hare A, Cannon GW, Griffiths MM, Wilder RL (1996) A genome scan localizes five non-MHC loci controlling collagen-induced arthritis in rats. *Nat Genet* 14: 82–85

54   Jirholt J, Cook A, Emahazion T, Sundvall M, Jansson L, Nordquist N, Pettersson U, Holmdahl R (1998) Genetic linkage analysis of collageninduced arthritis in the mouse. *Eur J Immunol* 28: 3321–3328

55   McIndoe RA, Bohlman B, Chi E, Schuster E, Lindhardt M, Hood L (1999) Localization of non-Mhc collagen-induced arthritis susceptibility loci in DBA/1j mice. *Proc Natl Acad Sci USA* 96: 2210–2214

56   Holmdahl R (2003) Dissection of the genetic complexity of arthritis using animal models. *J Autoimmun* 21: 99–103

57   Griffiths MM, Remmers EF (2001) Genetic analysis of collagen-induced arthritis in rats: a polygenic model for rheumatoid arthritis predicts a common framework of cross-species inflammatory/autoimmune disease loci. *Immunol Rev* 184: 172–183

58   Holmdahl R, Bockermann R, Jirholt J, Johansson A, Olofsson P, Lu S (2001) Elucidation of pathways leading to rheumatoid arthritis by genetic analysis of animal models. *Curr Dir Autoimmun* 3: 17–35

59   Brenner M, Meng HC, Yarlett NC, Griffiths MM, Remmers EF, Wilder RL, Gulko PS (2005) The non-major histocompatibility complex quantitative trait locus Cia10 contains a major arthritis gene and regulates disease severity, pannus formation, and joint damage. *Arthritis Rheum* 52: 322–332

60   Ribbhammar U, Flornes L, Backdahl L, Luthman H, Fossum S, Lorentzen JC (2003)

High resolution mapping of an arthritis susceptibility locus on rat chromosome 4, and characterization of regulated phenotypes. *Hum Mol Genet* 12: 2087–2096

61 Olofsson P, Holmberg J, Tordsson J, Lu S, Akerstrom B, Holmdahl R (2003) Positional identification of Ncf1 as a gene that regulates arthritis severity in rats. *Nat Genet* 33: 25–32

62 Hultqvist M, Olofsson P, Holmberg J, Backstrom BT, Tordsson J, Holmdahl R (2004) Enhanced autoimmunity, arthritis, and encephalomyelitis in mice with a reduced oxidative burst due to a mutation in the Ncf1 gene. *Proc Natl Acad Sci USA* 101: 12646–12651

63 Jarvinen P, Aho K (1994) Twin studies in rheumatic diseases. *Semin Arthritis Rheum* 24: 19–28

64 Gregersen PK (1999) Genetics of rheumatoid arthritis: confronting complexity. *Arthritis Res* 1: 37–44

65 Kremer JM, Phelps CT (1992) Long-term prospective study of the use of methotrexate in the treatment of rheumatoid arthritis. Update after a mean of 90 months. *Arthritis Rheum* 35: 138–145

66 Wolfe F, Michaud K, Stephenson B, Doyle J (2003) Toward a definition and method of assessment of treatment failure and treatment effectiveness: the case of leflunomide versus methotrexate. *J Rheumatol* 30: 1725–1732

67 Alarcon GS, Tracy IC, Blackburn WD Jr (1989) Methotrexate in rheumatoid arthritis. Toxic effects as the major factor in limiting long-term treatment. *Arthritis Rheum* 32: 671–676

68 McLeod HL, Evans WE (2001) Pharmacogenomics: unlocking the human genome for better drug therapy. *Annu Rev Pharmacol Toxicol* 41: 101–121

69 Gulcher J, Kong A, Stefansson K (2001) The genealogic approach to human genetics of disease. *Cancer J* 7: 61–68

70 Grant SF, Thorleifsson G, Frigge ML, Thorsteinsson J, Gunnlaugsdottir B, Geirsson AJ, Gudmundsson M, Vikingsson A, Erlendsson K, Valsson J et al (2001) The inheritance of rheumatoid arthritis in Iceland. *Arthritis Rheum* 44: 2247–2254

71 Lin JP, Hirsch R, Jacobsson LT, Scott WW, Ma LD, Pillemer SR, Knowler WC, Kastner DL, Bale SJ (1999) Genealogy construction in a historically isolated population: application to genetic studies of rheumatoid arthritis in the Pima Indian. *Genet Med* 1: 187–193

72 Mokwe E, Ohmit SE, Nasser SA, Shafi T, Saunders E, Crook E, Dudley A, Flack JM (2004) Determinants of blood pressure response to quinapril in black and white hypertensive patients: the Quinapril Titration Interval Management Evaluation trial. *Hypertension* 43: 1202–1207

Index

The PIR-Series
Progress in Inflammation Research

Homepage: http://www.birkhauser.ch

Up-to-date information on the latest developments in the pathology, mechanisms and therapy of inflammatory disease are provided in this monograph series. Areas covered include vascular responses, skin inflammation, pain, neuroinflammation, arthritis cartilage and bone, airways inflammation and asthma, allergy, cytokines and inflammatory mediators, cell signalling, and recent advances in drug therapy. Each volume is edited by acknowledged experts providing succinct overviews on specific topics intended to inform and explain. The series is of interest to academic and industrial biomedical researchers, drug development personnel and rheumatologists, allergists, pathologists, dermatologists and other clinicians requiring regular scientific updates.

Available volumes:
T Cells in Arthritis, P. Miossec, W. van den Berg, G. Firestein (Editors), 1998
Chemokines and Skin, E. Kownatzki, J. Norgauer (Editors), 1998
Medicinal Fatty Acids, J. Kremer (Editor), 1998
Inducible Enzymes in the Inflammatory Response,
    D.A. Willoughby, A. Tomlinson (Editors), 1999
Cytokines in Severe Sepsis and Septic Shock, H. Redl, G. Schlag (Editors), 1999
Fatty Acids and Inflammatory Skin Diseases, J.-M. Schröder (Editor), 1999
Immunomodulatory Agents from Plants, H. Wagner (Editor), 1999
Cytokines and Pain, L. Watkins, S. Maier (Editors), 1999
In Vivo *Models of Inflammation*, D. Morgan, L. Marshall (Editors), 1999
Pain and Neurogenic Inflammation, S.D. Brain, P. Moore (Editors), 1999
Anti-Inflammatory Drugs in Asthma, A.P. Sampson, M.K. Church (Editors), 1999
Novel Inhibitors of Leukotrienes, G. Folco, B. Samuelsson, R.C. Murphy (Editors), 1999
Vascular Adhesion Molecules and Inflammation, J.D. Pearson (Editor), 1999
Metalloproteinases as Targets for Anti-Inflammatory Drugs,
    K.M.K. Bottomley, D. Bradshaw, J.S. Nixon (Editors), 1999
Free Radicals and Inflammation, P.G. Winyard, D.R. Blake, C.H. Evans (Editors), 1999
Gene Therapy in Inflammatory Diseases, C.H. Evans, P. Robbins (Editors), 2000
New Cytokines as Potential Drugs, S. K. Narula, R. Coffmann (Editors), 2000
High Throughput Screening for Novel Anti-inflammatories, M. Kahn (Editor), 2000
Immunology and Drug Therapy of Atopic Skin Diseases,
    C.A.F. Bruijnzeel-Komen, E.F. Knol (Editors), 2000
Novel Cytokine Inhibitors, G.A. Higgs, B. Henderson (Editors), 2000
Inflammatory Processes. Molecular Mechanisms and Therapeutic Opportunities,
    L.G. Letts, D.W. Morgan (Editors), 2000